Bio-Inspired Engineering

Bio-Inspired Engineering

C. H. JENKINS

Montana State University, Bozeman, MT 59717

MOMENTUM PRESS

MOMENTUM PRESS, LLC, NEW YORK

Bio-Inspired Engineering

Copyright © Momentum Press®, LLC, 2012

All rights reserved. No part of this publication may be reproduced, stored in a retrieval system, or transmitted in any form or by any means—electronic, mechanical, photocopy, recording or any other—except for brief quotations, not to exceed 400 words, without the prior permission of the publisher.

First published by
Momentum Press®, LLC
222 East 46th Street
New York, NY 10017
www.momentumpress.net

ISBN-13: 978-1-60650-223-5 (hard back, case bound)
ISBN-10: 1-60650-223-9 (hard back, case bound)

ISBN-13: 978-1-60650-225-9 (e-book)
ISBN-10: 1-60650-225-5 (e-book)

DOI: 10.5643/9781606502259

Cover design by Jonathan Pennell
Interior design by MPS

10 9 8 7 6 5 4 3 2

Printed in Canada

To my students, who have taught me many things.

Table of Contents

Preface	xv
1. Introduction	1
1.1 A Classic Story of Bio-inspired Engineering	1
1.2 Bio-inspired Engineering Defined	2
1.3 BiE Yesterday and Today	4
1.4 Biology vs. Engineering	6
Key Points to Remember	7
References	7
Problems	7
2. Design	9
2.1 Introduction	9
2.2 Engineering Design	9
2.3 Design in Nature	18
Key Points to Remember	23
References	23
Problems	24
3. Structures and Materials in Engineering	27
3.1 Overview of Structures	27
3.2 Application to Engineering Structures	32
3.3 Constitutive Models	41
Key Points to Remember	52
References	52
Problems	53
4. Structures and Materials in Nature	57
4.1 Configuration in Nature	57
4.2 Shape and Structure in Nature	63
4.3 Natural Materials	77
Key Points to Remember	82
References	82
Problems	82

Table of Contents

5. Compliant Structures	**85**
5.1 The Compliant Paradigm in Nature	85
5.2 Examples of Compliant Structures in Nature and Engineering	87
5.3 Compliant Structure Response Analysis	91
Key Points to Remember	104
References	104
Problems	104
6. Smart Structures	**107**
6.1 Introduction	107
6.2 Biological Smartness	109
6.3 The Biological Neuron	110
6.4 Self-Healing in Biology and Engineering	111
Key Points to Remember	113
References	114
Problems	114
7. The Bio-inspired Engineering Process	**117**
7.1 Forward and Reverse Design Redux	117
7.2 Resources for Bio-inspired Engineering	119
7.3 Reverse Design for Bio-inspired Engineering	122
Key Points to Remember	126
References	126
Problems	127
Appendices	
Appendix 1 Biological Classification	131
Appendix 2 A Simple Inverse Mechanics Problem	133
Appendix 3 Reverse Engineering Worksheets	135
Appendix 4 Tip Angle θ_0 and the Elliptic Integral of the First Kind $K(\xi)$	141
Appendix 5 Comparative Properties of Natural Materials	143
Figure Attributions	**147**
Index	**151**

List of Color Plates

Figure 2.9. Mangrove roots. Roots play both structural and mass transport roles and are structured accordingly (HD). Redundant load paths (roots) allow for changing conditions (JGE). The roots live in an environment that would challenge many engineered structures. http://commons.wikimedia.org/wiki/File:Mangrove.jpg.

Figure 2.10a–b. Wings. Different elements support tensile loads, compressive loads, or control airflow (PSF). (a) Flexibility gives maneuverability and ability to change shape as needed (LM). http://en.wikipedia.org/wiki/File:Big-eared-townsend-fledermaus.jpg. (b) Many feathers give damage tolerance and can be replaced individually (PSF). http://commons.wikimedia.org/wiki/File:Green_Woodpecker_wing.jpg.

Figure 4.13. Material structural hierarchy in the root of the tritium plant.

Figure 5.4. Fry of the paradise fish or paradise gouramis (*Macropodus opercularis*) in their bubble nest.

Figure 5.6. The bullfrog's throat is an inflatable membrane.

Figure 5.9. Scientific high-altitude balloon.

Figure 5.10. The 20 m ATK solar sail undergoing deployment testing in the NASA Plum Brook vacuum chamber.

Figure 6.4. Sap infusing into a tree's wound.

Figure 6.5. Phases of wound healing.

Figure E.7.1. Three-horned rhinoceros beetle.

Figure E.7.2. Concept of possible new product inspired by the three-horned rhinoceros beetle.

List of Figures

Figure 1.1.	Close-up views of (a) burr hooks and (b) Velcro.	2
Figure 1.2.	Technical drawing from the original Velcro patent of George De Mestral depicting hook fabrication (above) and joined system (below).	4
Figure 1.3.	(a) Rope in biology (vines) and (b) Ancient Egyptian artwork depicting the manufacture of rope by mechanical means.	4
Figure 1.4.	Erected shelter: (a) the nest of the paper wasp and (b) African hut.	5
Figure 1.5.	Sailing in (a) biology and (b) early technology (Kon Tiki shown here).	5
Figure 1.6.	Examples of Leonardo da Vinci's inspiration from biological flight in ca. 1490: (a) study of a wing structure and (b) flying machine.	6
Figure 2.1.	Forward vs. inverse design. BiE expands the solution space.	10
Figure 2.2.	Physical coordinate space.	11
Figure 2.3.	The solid/liquid/gas phase space diagram for water.	12
Figure 2.4.	Abstract design space.	12
Figure 2.5.	Design space constrained by limits on one of the DDOF.	13
Figure 2.6.	A generic constrained optimal design problem.	14
Figure E.2.1.	A generic rectangular tray. Dimensions a, b, and c are internal dimensions.	14
Figure E.2.2.	Plot of Q vs. b.	15
Figure E.2.3.	The meaning of b_{opt}.	16
Figure E.2.4.	Graph for Example 2.2.	17
Figure 2.7.	Sectioned halves of parts brushes: (a) parts brushes without flow channel and (b) parts brushes with flow channel. In (a) and (b), Company A's brush is the pair on the left, and Company B's brushes are the middle and rightmost pairs.	18
Figure 2.8.	Spider web. Different varieties of threads have different capabilities and functions (PSF and HD). Redundant load paths (threads) provide damage tolerance (JGE). Flexibility absorbs impacts and shares load equally (LM).	20
Figure 2.9.	Mangrove roots. Roots play both structural and mass transport roles and are structured accordingly (HD). Redundant load paths (roots) allow for changing conditions (JGE). The roots live in an environment that would challenge many engineered structures.	21

List of Figures

Figure 2.10.	Wings. Different elements support tensile loads, compressive loads, or control airflow (PSF). (a) Flexibility gives maneuverability and ability to change shape as needed (LM). (b) Many feathers give damage tolerance and can be replaced individually (PSF).	21
Figure 2.11.	Palm trees. Leaves and trunk are flexible and help to handle wind loads (LM). Trunk is a multifunction structure (HD).	22
Figure 3.1.	Simple load orientations. Single-headed arrows are forces; double-headed arrows are moments (that follow the right-hand rule).	28
Figure 3.2.	Definition sketch of an axial structure.	32
Figure 3.3.	Torsion structure of constant circular cross section. Any point on the circular cross section x = constant can be identified by the cylindrical–polar coordinates (r, θ, x).	33
Figure 3.4.	A moment being applied to a shaft by a force P on a wrench of length d.	34
Figure 3.5.	Global free-body diagram of an axial structure.	34
Figure 3.6.	Local free-body diagram of an axial structural element.	35
Figure 3.7.	Internal force system on a differential shaft element consisting of a shear force ΔF.	36
Figure E.3.1.	(a) Configuration; (b) global free-body diagram; (c) local FBD; (d) stress variation along the structure length.	37
Figure 3.8.	Differential element of radius r coaxial with shaft of radius R with sector $r\Delta\theta$ shown.	40
Figure 3.9.	Differential elements of the shaft under the action of a torque M: (a) Isometric view and (b) plan view. Note that the originally rectangular differential element $r\Delta\theta$ by Δx has been <u>sheared</u> into a parallelogram.	40
Figure 3.10.	Loading–unloading curve for a linear elastic material.	42
Figure 3.11.	Nonlinear elastic material loading–unloading curve.	43
Figure 3.12.	The concept and definition of stretch.	44
Figure 3.13.	Range of material behavior.	45
Figure 3.14.	The three stages of creep.	46
Figure 3.15.	Creep and recovery: (a) load history and (b) strain history.	46
Figure 3.16.	Stress relaxation: (a) load history and (b) stress history.	47
Figure 3.17.	A viscous dashpot.	47
Figure 3.18.	A viscoelastic solid model.	48
Figure 3.19.	Creep response for the Kelvin–Voigt Solid.	49
Figure 3.20.	Spring and damper connected in series.	49
Figure 3.21.	Stress relaxation in the Maxwell model.	50
Figure 3.22.	The concept of the Standard Linear Solid.	50
Figure 3.23.	Plot of the Standard Linear Solid creep response.	51
Figure 4.1.	Geometric scaling. Doubling the length increases the area by four times.	58

List of Figures

Figure 4.2.	Log–log plot of equation (4.1). The function $y = \sqrt{x}$ has a slope of ½ (e.g., $\log(10/1)/\log(100/1) = ½$). The function $y = x^{1/5}$ has a slope of ⅕, etc. (Verification of these slopes is left as an exercise for the student.)	59
Figure E.4.1.	Log–log plot of Table E4.1.	61
Figure 4.3.	Geometry of the Koch snowflake. This geometry has fractal dimension of 1.26. (a) Begin with an equilateral triangle. (b) Replace the middle third of each line segment with a pair of line segments forming an equilateral "bump." (c) Repeat.	62
Figure 4.4.	Fractal nature of tree branching.	63
Figure 4.5.	Bifurcation branching with cylinders.	63
Figure E.4.2.	A plot of A_s vs. R for a closed right circular cylinder.	65
Figure E.4.3a.	Tapered shaft definition sketch.	68
Figure E.4.3b.	Non-dimensional shear stress vs. shaft length	69
Figure 4.6.	Plant stem of thin circular closed cross section. The mean radius is given by r_m, and h is the wall thickness.	70
Figure E.4.4.	Shaft of thin square closed cross section. A side has a mean length given by b and a wall thickness h.	71
Figure 4.7.	A longitudinal cut transforms a closed cross section into an open cross section.	72
Figure 4.8.	Cross section of a thin solid rectangular shaft.	72
Figure 4.9.	Approximate torsion constants for some thin-walled open sections: (a) angle, (b) channel, and (c) circular arc.	73
Figure E.4.5.	Circular beam cross section.	76
Figure 4.10.	The hierarchical organization.	78
Figure 4.11.	A hierarchy in systems biology.	79
Figure 4.12.	Material structural hierarchy in the root of the tritium plant.	80
Figure 5.1.	A toy balloon is a quintessential membrane structure.	86
Figure 5.2.	The significance of shape change on load-carrying capacity. (a) A thin sheet of paper can barely support its own weight in cantilever configuration. (b) With a slight initial curvature added, the sheet is now capable of carrying a small additional load. (c) A significant initial shape change allows the same sheet to bear a much greater load.	86
Figure 5.3.	Blood cells. In this scanning electron microscope image, you can see red blood cells, several white blood cells including lymphocytes, and many small disk-shaped platelets.	87
Figure 5.4.	Fry of the paradise fish or paradise gouramis (*Macropodus opercularis*) in their bubble nest.	87
Figure 5.5.	Blades of grass are slender compliant columns.	88
Figure 5.6.	The bullfrog's throat is an inflatable membrane.	88
Figure 5.7.	The City of Bridgetown's (Barbados) Grantley Adams International Airport. The translucent membrane roof provides pleasant shelter between the departure and arrival halls.	89

List of Figures

Figure 5.8.	A modern ram air parasail.	89
Figure 5.9.	Scientific high-altitude balloon.	90
Figure 5.10.	The 20-m ATK solar sail undergoing deployment testing in the NASA Plum Brook vacuum chamber.	90
Figure 5.11.	Euler column buckling: (a) initial configuration; (b) deformed configuration (deformation shown greatly exaggerated); (c) free-body diagram.	91
Figure 5.12.	Compliant column definition sketch.	93
Figure 5.13.	Definition sketch of a tip-loaded prismatic cantilever beam shown in the deformed configuration. The beam is represented by its neutral axis. At the lower right is a free-body diagram (FBD).	96
Figure 5.14.	Compliant chain definition sketch. In the figure, ds is an increment of arc length along the chain.	100
Figure 5.15.	Close up view of a link chain.	100
Figure E.5.1.	Stiff and compliant hanging cables.	103
Figure 6.1.	The range of smart structure complexity: (a) bimetallic thermostat; (b) laser-guided automated guided vehicle; (c) a flapping wing microair vehicle as proposed by the US Air Force, shown in a computer simulation; (d) self-healing structure.	108
Figure 6.2.	The bimetallic thermostat.	109
Figure 6.3.	The biological neuron.	111
Figure 6.4.	Sap infusing into a tree's wound.	112
Figure 6.5.	Phases of wound healing.	113
Figure 6.6.	A resin infusion self-healing engineered system.	113
Figure 7.1.	The desired forward design process in bio-inspired engineering.	118
Figure E.7.1.	Three-horned rhinoceros beetle.	123
Figure E.7.2.	Concept of possible new product inspired by the three-horned rhinoceros beetle.	126
Figure A.1.	The seven main ranks in biological classification.	131
Figure A.2.1.	An SDOF spring–mass–damper system. Here, k = spring stiffness, m = mass, and η = damping coefficient. This system is also called a *harmonic oscillator*.	133
Figure A.2.2.	Nondimensional $X(\omega)$ as function of $r = \omega/\omega_n$.	134
Figure A.5.1.	Modulus vs. density. Reproduced with permission from Ashby, M. F. 2005. *Materials Selection in Mechanical Design*, 3rd ed., 395–397. Elsevier.	143
Figure A.5.2.	Strength vs. density. Reproduced with permission from Ashby, M. F. 2005. *Materials Selection in Mechanical Design*, 3rd ed., 395–397. Elsevier.	144
Figure A5.3.	Modulus vs. strength. Reproduced with permission from Ashby, M. F. 2005. *Materials Selection in Mechanical Design*, 3rd ed., 395–397. Elsevier.	145

Preface

> But nature is always more subtle, more intricate, more elegant than what we are able to imagine.
>
> —*Carl Sagan*

This text is a small, humble step toward putting the power of bio-inspiration into the hands of engineers and others who are designing our technological world. Hopefully through this book and the other good work going on in the field today, we can begin to close the gap between the elegant solutions of nature and the possibilities we can imagine.

The primary purpose of this book, and of bio-inspired engineering (BiE) in general, is to expand the design space of possible solutions to technical problems. We do that of course by understanding how nature solves problems, which starts with admitting that nature has something to teach us. Additionally and at least as important is knowing how to translate biological knowledge into engineering practice. Even though the domain of biology is vast and new discoveries occur daily, much is known about biological solutions. Turning this knowledge into technical solutions is the challenge we face—it is also the focus of considerable attention in modern BiE, and hence of this book as well.

This book has a decidedly engineering flavor. It is not a book on biology for engineers; it is, however, a book for technologists looking for a portal to the biological world. Engineers and engineering students should find comfortable footing from which to leap, and biologists will find plenty of opportunities filling in what is missing! There are a wealth of topics that could be covered in a BiE text (and that will be covered someday no doubt!), but the topics chosen here follow the sage advice to all authors—write what you know.

The present text is a natural evolution of the author's work with compliant and smart structures for nearly three decades, and from teaching two courses in particular: an undergraduate course ME 455 Bio-Inspired Engineering and a graduate course ME 555 Smart Structures at Montana State University. The approach taken in this book is one honed from those years of experience—we can access biology through our engineering knowledge. For example, in Chapter 2 we learn about design in nature by first reviewing what we know about engineering design. One of the essential concepts we encounter is that of optimum design. Optimum design in engineering is a reasonable mature subject and we provide a brief review. Optimality in biology, however, is the subject of current debate within the field itself and based on our engineering knowledge we can appreciate and learn from that debate.

The same approach is used elsewhere in the text. Chapter 3 reviews structure and materials in engineering before exploring those topics in nature in Chapter 4. Chapter 5 deals with the ubiquitous case of highly flexible or compliant structures in nature by first studying these structures in engineering. Chapter 6 on smart structures follows suit. In Chapter 7, we try to provide guidance on how to actually *do* bio-inspired engineering.

Biological solutions do not exist in isolation but form part of an interdependent, interconnected whole, and the same is true for this book. *Bio-Inspired Engineering* would not exist if it weren't for the help and contribution of many others. First and foremost are my students, those who have worked with me on research and those who have suffered through my classes. Having put up with this writing exercise on many prior occasions, my wife deserves sainthood for her continued patience with me; that she agreed to do much of the typing can only be explained by love. Finally, Joel Stein and Millicent Treloar at Momentum Press provided much appreciated support and the never ending extensions.

Go forth and close the gap.

Chris Jenkins
Bozeman, MT

CHAPTER 1

Introduction

If any of your employees ask for a two-week holiday to go hunting, say yes.

—*George De Mestral*

We will shortly begin our study of bio-inspired engineering by examining one of the classic examples in the field. We will first meet George De Mestral and learn of his inspiration from biology for a technology that is very common today. This chapter will provide the student with an introduction to bio-inspired engineering through examples, definitions, and exercises.

1.1 A CLASSIC STORY OF BIO-INSPIRED ENGINEERING

George De Mestral was born near Lausanne, Switzerland, on June 19, 1907. Inventive as a child, De Mestral went on to study engineering at the famous Ecole Polytechnique in Lausanne. Many years later, in the 1940s, De Mestral was hunting with his dog in the mountains of Switzerland. He became curious about the burrs (apparently burdock seeds, similar to cockleburs) stuck to his clothing and to his dog's fur. Examining the seeds later under a microscope, he discovered nature's design for this method of seed dispersal—hitchhiking! The tips of the burr whiskers are formed like hooks so they can easily catch on the fur of animals that brush against the plant, while also inconveniently catching on fabric loops in clothing (see Figure 1.1). De Mestral wondered if there wasn't some way to use nature's idea in the fastening and joining of human artifacts.

George De Mestral labored for many years to move from his initial encounter with natural design to his commercially viable "Velcro"® fastener (from the French words *velour*, meaning velvet, and *crochet*, meaning hook). Although a number of engineering developments were ultimately required, particularly in the areas of materials and

1

manufacturing, he filed for a patent in 1952 that was finally granted in 1955. De Mestral envisioned Velcro as a replacement for the ubiquitous zipper, finally making significant inroads with NASA through use in astronaut space suits. Figure 1.1 shows a close-up view of the modern Velcro "hook-and-loop" fastener. We'll come back to De Mestral's outstanding invention shortly after we discuss some terminology.

Figure 1.1. Close-up views of (a) burr hooks (http://commons.wikimedia.org/wiki/File:Big_Burrs.jpg) and (b) Velcro (http://en.wikipedia.org/wiki/File:Velcro.jpg).

1.2 BIO-INSPIRED ENGINEERING DEFINED

As you come to study the subject, you will encounter several different words that are used to describe technology derived from knowledge about biological solutions. The term "bionic" was introduced in 1960 to describe the interest shown by the US

Air Force in how biological design might inform engineering design and technology (Vincent et al., 2006). Bionic may be loosely translated from a Greek word meaning "like life." Historically, however, bionics has been more or less associated with biomedical engineering applications such as prostheses. During the same decade, the term "biomimetics" was also introduced. Biomimetics and its variants such as biomimicry derive from *bios*, meaning life, and *mimesis*, meaning to imitate. (Biomimetics was apparently first used by Otto Schmitt in 1969 (Vincent et al., 2006).) According to the Biomimicry Institute (biomimicryinstitute.org): "Biomimicry is a new discipline that studies nature's best ideas and then imitates these designs and processes to solve human problems." "Biomimicry—the conscious emulation of life's genius" (Benyus, 1997). The institute also emphasizes the importance of sustainability in biomimicry, that "Biomimicry is a new science that studies nature's models and then emulates these forms, process, systems, and strategies to solve human problems—sustainably."

In this text, we use the term "bio-inspired engineering" or BiE for short. BiE means literally to take inspiration from nature's solutions as a basis for solving technical problems. If we don't take any of these various terms too literally, we see that they all are trying to describe the same basic idea. We prefer "inspiration" over "mimicking" or "imitating," however, due to the extreme challenge of imitating nature. We will come to see that many natural solutions are so highly complex that their imitation would be daunting. Sustainability, while absolutely the right thing, can be very difficult to assess, given the complex interconnectedness of technology and life, and the time span for change to occur. Inspiration is a much more accessible, and still laudable, goal, and for the sake of clarity we take that as the core of our definition.

So now let's return to De Mestral and reconsider his invention in the light of our understanding thus far. We know of his inspiration resulting from the burrs stuck to himself and his dog during their hike in the mountains. We know of his further investigation of the adhesion mechanism of the burrs, including microscopy. We know of his patent application. Figure 1.2 shows technical drawings from the application.

De Mestral was led finally to a process of using synthetic fibers (the emergent Nylon) in a weave that formed large loops (6 in Figure 1.2) that were cut to form the hooks (4), leaving a remnant of the original loop (10). Is this mimicking nature? If it is, what are the essential natural features being emulated? Certainly essential features of the burr adhesion mechanism must include the fibers extending outward from the seed, the hook-like termination of the fiber, and the fiber stiffness and strength. Based upon these features, it could be readily argued that the engineered artifact does indeed mimic the burr. But are there other features that are not replicated? Look closely at Figure 1.1a. Note the roughly ellipsoidal shape of the burr and the associated uniform distribution of hooks that allow the burr to catch from any orientation. De Mestral's patent (Velcro, 1955) describes a planar configuration that will only catch in one orientation (not dissimilar to many modern Velcro products). Is it sustainable? This is difficult to answer; the fastener uses petroleum-based polymers such as Nylon and requires energy to manufacture. Does it replace even more resource-consuming processes such that a net reduction in required resources is the result? What about end of life—is the product recyclable or biodegradable? These are the kinds of issues that must be considered to make judgments on mimicking and sustainability. Again, in this text we will focus primarily on biological inspiration.

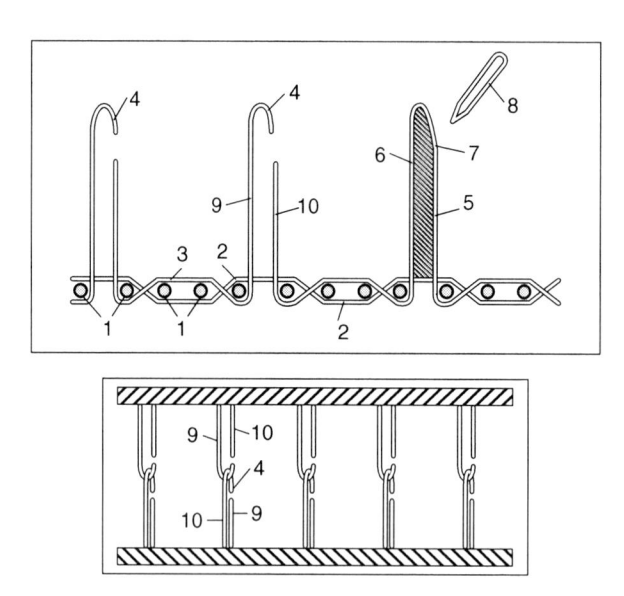

Figure 1.2. Technical drawing from the original Velcro patent of George De Mestral depicting hook fabrication (above) and joined system (below).

1.3 BiE YESTERDAY AND TODAY

It is a reasonable assertion that human designers, engineers, technologists, and the like have always taken cues from nature to some extent. For the earliest technology developments, biological inspiration is even more likely, given the intimate connection between humans and their environment before the advent of cities, industry, etc. For example, imagine the development of rope. The earliest ropes were probably lengths of vines (Figure 1.3a) and other fibrous plant stalks, first singly, then in braided strands. Human-made rope seems closely linked with the early rope inspired by biology (Figure 1.3b).

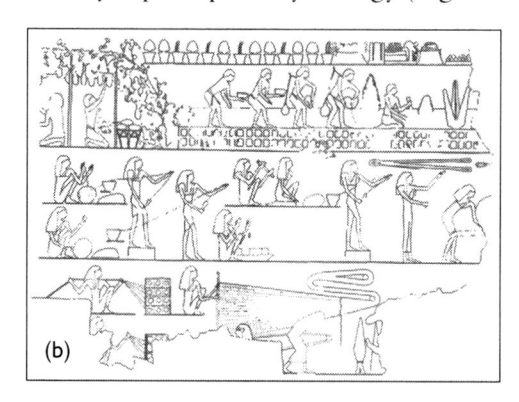

Figure 1.3. (a) Rope in biology (vines) (http://commons.wikimedia.org/wiki/File:Dead_ivy_vines_cling_to_tree.jpg) and (b) Ancient Egyptian artwork depicting the manufacture of rope by mechanical means (http://en.wikipedia.org/wiki/File:Ancient_Egypt_rope_manufacture.jpg).

Shelter was obviously very important to early humans. Given that many early peoples were nomadic, lightweight and deployable shelter, or shelter readily erectable from local materials, would have been desirable. Looking at the natural world, examples of lightweight and rapidly erectable shelter can be found, such as in the nest of the paper wasp (Figure 1.4a). It is not hard to imagine early nomads learning from the natural world (Figure 1.4b).

Figure 1.4. Erected shelter: (a) the nest of the paper wasp (http://commons.wikimedia.org/wiki/File:Wasp_March_2008-9.jpg) and (b) African hut (http://commons.wikimedia.org/wiki/File:Case_%C3%A0_la_chefferie_de_Bana.jpg).

Transportation was another driver of early technology development. Sailing was an early and powerful form of transportation. An example of sailing existing in nature is the man-of-war that uses an inflatable bladder as a sail (Figure 1.5a). Early human-designed sailcraft was little more than a single small sail connected to a mast and could easily have been derived from natural solutions (Figure 1.5b).

Figure 1.5. Sailing in (a) biology (http://en.wikipedia.org/wiki/File:Portuguese_Man_O%27_War_Miami_March_2008.jpg) and (b) early technology (Kon Tiki shown here) (http://en.wikipedia.org/wiki/File:Kon-Tiki.jpg).

Certainly by the time of da Vinci (1452–1519), we had a clear record of bio-inspired engineering (Figure 1.6). More recently, engineers have been looking to biology for solutions in robotics, stronger materials, self-healing structures and systems, sensors, and many other important areas. We will cover many of these developments later in this text.

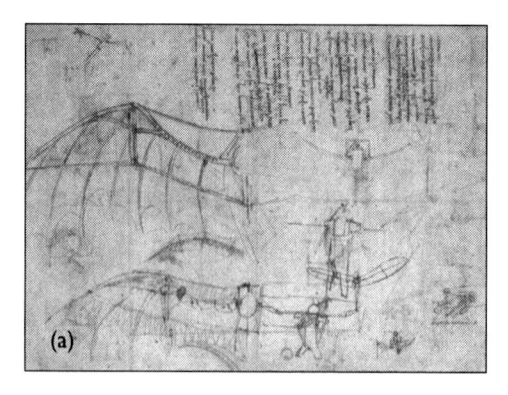

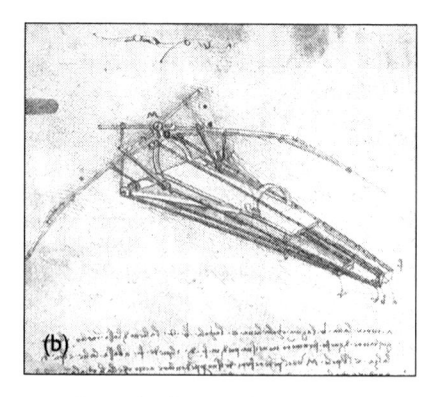

Figure 1.6. Examples of Leonardo da Vinci's inspiration from biological flight in ca. 1490: (a) study of a wing structure (http://commons.wikimedia.org/wiki/File:Leonardo_Design_for_a_Flying_Machine,_c._1488.jpg) and (b) flying machine (http://en.wikipedia.org/wiki/File:Design_for_a_Flying_Machine.jpg).

1.4 BIOLOGY VS. ENGINEERING

When asked about the difference between science and engineering, the famous engineer Theodore von Karman said that "Scientists discover what is, engineers create what never was." Biology is a largely observational (empirical) field of study that tries to understand and describe the living world. That challenging goal is further complicated by the vast length and time scales over which biology operates, from the nanolevel to global scale and from microseconds to the time scale of evolution. The biologist tries to understand and interpret a very complex story. (See Appendix 1 for the biological classification system.)

In contrast, engineering sits at the intersection of science and art. The engineer uses knowledge of mathematics and science in a creative process to develop new solutions for the benefit of society. This process is called "design," and we will talk more about design in the next chapter. The engineer's job is to make something new that satisfies (hopefully optimally) the needs of a customer. Creating something new, which doesn't exist, is a challenging and humbling task but one that itself has obvious parallels with nature.

One issue that arises when two disparate disciplines interact is language. The same word may have entirely different meanings in two different disciplines. For example, consider the word "stress." To the biologist, the first meaning that comes to mind for stress would be "environmental stress" or "psychological stress"—that is, when a biological organism is presented with a threat (emotional, environmental, physical) and fails to respond appropriately or successfully. To the engineer, stress is simply the internal response of a material to applied loads, strictly the internal force developed in response to an applied load normalized by the area over which the force acts. Bridging the language divide between biology and engineering will take constant vigilance on our part.

KEY POINTS TO REMEMBER

- Keep in mind the subtle differences between biomimetics and bio-inspired.
- Bio-inspired engineering is undergoing a modernization, but human technologists have likely always taken their cue from nature.
- Different disciplines have different meanings for the same words, and this can be very challenging in the BiE process.

REFERENCES

Benyus, J. M. 1997. *Biomimicry.* William Morrow & Co.

http://biomimicryinstitute.org/about-us/what-is-biomimicry.html.

Velcro, S. A. 1955. *Improvements in or Relating to a Method and a Device for Producing Velvet Type Fabric.* Switzerland Patent no. 721338.

Vincent, F. V., O. A. Bogatyreva, N. R. Bogatyrev, A. Bowyer, and A.-K. Pahl. 2006. "Biomimetics: Its Practice and Theory." *J. Royal Society Interface* 3:471–482. doi: 10.1098/rsif.2006.0127; http://rsif.royalsocietypublishing.org/content/3/9/471.

PROBLEMS

1.1. In this chapter, we have discussed some of the salient features of burrs that relate to their "hook-and-catch" capability and how those features may or may not map to features found in Velcro. Study the burr more closely for other hook-and-catch features and discuss how these may or may not map to features found in Velcro. If you conclude a feature doesn't map, explore concepts to enable a mapping.

1.2. Investigate other recent examples of bio-inspired engineering:

　a. Adhesive inspired by the feet of geckos

　b. Self-cleaning surfaces inspired by the leaves of the lotus plant

　c. Antireflective surfaces inspired by insect eyes and wings

　d. Sailboat hulls inspired by the skin of sharks

1.3. Investigate other recent examples of bio-inspired engineering:

　a. Architecture inspired by biological membrane structures

　b. Composite materials inspired by biological composites such as bamboo

c. Control systems inspired by neural circuits

d. Aircraft wing improvements inspired by bird wings

e. Navigation systems inspired by bats, bees, and others

1.4. For each of the common engineering terms below, discuss the meaning (if any) of the same word in biology:

a. Strain

b. Structure

c. Function

d. Matrix

1.5. For each of the common biological terms below, discuss the meaning (if any) of the same word in engineering:

a. Evolution

b. Equilibrium

c. Membrane

d. Fiber

1.6. What is life? Describe the characteristics of living systems.

1.7. Projects: Investigate early bio-inspired

a. flight

b. shelter

c. sailing

d. tools (e.g., rope)

CHAPTER 2

Design

Biology is engineering.

—*Daniel Dennett*

Design is the essence of engineering. In this chapter, we introduce the basic concepts in engineering design, such as design process, design space, design rules, optimal design, and reverse engineering. As our goal is to "expand the design solution space" through a knowledge of how nature solves problems, in this chapter we try to come to some understanding about how nature "designs."

2.1 INTRODUCTION

Design is a creative human endeavor that is the essence of engineering. Design sits at the intersection of science, art, business, politics, and psychology, and in this chapter we will study the process of design in brief. The concept and practice of *design* are human constructs. We will explore the idea of *design in nature* in this chapter, but we keep in mind the essential difference between human and natural designs. The central reason for the study of BiE is to expand the design space of possible engineering solutions.

2.2 ENGINEERING DESIGN

2.2.1 The Engineering Design Process

Design is an activity that turns the abstract information about the <u>need</u> for a product or a process into concrete knowledge necessary for <u>realizing</u> that product or process. Design is an iterative activity that proceeds through various stages, going from the abstract to the concrete by successive refinement. Different authors or organizations will define the design process differently, yet those differences are usually slight and there is generally good agreement overall about what constitutes the design process (Dieter, 2000; Dym and Little, 2004; French, 1988; Shigley and Mischke, 2001; Ullman, 1997). It is imperative that the design process be "total design." Total design encompasses *life cycle* design, in that the entire product life cycle is considered, from "cradle to grave." For example, in the total design of a product, considerations should be given to

1. extracting and processing of materials;
2. prediction of response;
3. manufacturing, assembling, and maintenance of the product;
4. removal of the product from service;
5. minimizing the environmental impact.

Specifically, resource cost of, and unintended consequences stemming from, these considerations should be made. (For the specific application to *total structural design*, see Jenkins and Khanna (2005).)

The design process as usually constructed supports the "forward design problem" (see Figure 2.1). Forward design seeks a solution to a set of requirements and constraints and represents the vast majority of design activity. The need and requirements drive the design solution in a logical fashion to higher levels of specificity. On the other hand, "inverse design" (or "reverse design") is knowledge driven: one takes a known solution and searches for a problem to solve with that solution (Figure 2.1). A good example of this inverse design is the case of Velcro described in Chapter 1. In fact, the inverse approach is quite common in BiE, and we will discuss more about these approaches in the

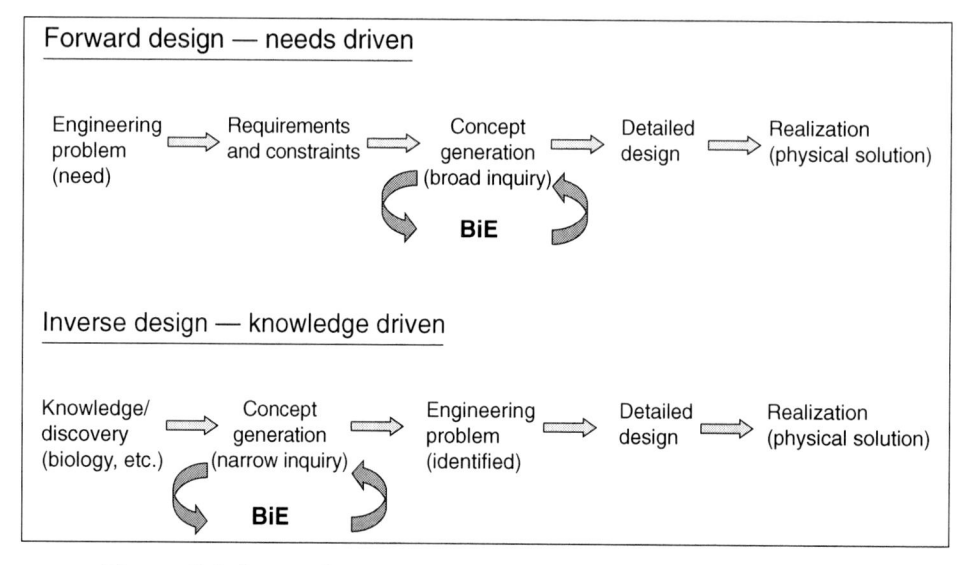

Figure 2.1. Forward vs. inverse design. BiE expands the solution space.

last chapter. (In engineering mechanics, e.g., the "forward problem" seeks the response of a system with a set of known parameters, say, the stress in a structural member, given the loads, boundary conditions, geometry, and material properties. The "inverse problem" seeks the identification of system parameters from a given response—this is particularly challenging, given that more than one set of parameters could result in the same response. See Appendix 2 for a simple example of an inverse mechanics problem.)

Design solutions are nonunique in the sense that different engineers will arrive at different design solutions; some may be better solutions than others (e.g., some closer to optimality than others), but there is no single, absolute solution to any design problem. This last point is important for us because it leads us to define the *design solution space*. Before we describe this abstract space, however, let's step back and think about physical space, for example, the space of the environment you're in right now. We usually define three-dimensional physical space by a set of three coordinates, say *x–y–z*. If you're indoors, there's probably an intersection of two walls and the floor that could quite nicely form the *coordinate axes* of the space (Figure 2.2).

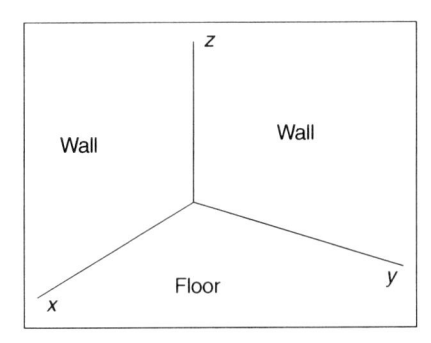

Figure 2.2. Physical coordinate space.

We can now think of the physical space as being defined by the coordinates and consequently define a variety of solutions in this space, e.g., the trajectory of a particle to given excitation. If that trajectory depends on each of the coordinates *independently*, then we would say the particle had three *coordinate degrees of freedom* (CDOF). A simple example would be the magnitude of the particle's position d from the origin:

$$d(x, y, z) = \sqrt{x^2 + y^2 + z^2} \tag{2.1}$$

If the particle was constrained to move only along the *x–z* wall, then there would be just two CDOF, since one of the coordinates is *constrained* by the equation $y = 0$, that is,

$$d(x, z) = \sqrt{x^2 + z^2}, \quad y = 0 \tag{2.2}$$

If the particle is now constrained to move in a circle of fixed radius R along the same wall, located with its center at (x_0, z_0), how many degrees of freedom are there? Well, there is only one CDOF, since x and z are no longer *independent*, but related by the equation of the circle

$$x^2 + z^2 = R^2 \tag{2.3}$$

(The position in this case will be left as an exercise for the student.)

A set of coordinates can thus define a *space*, and this can be a physical space as above or an abstract space, like the complex space of real and imaginary numbers. In a like manner, we will use a variety of quantities as *generalized coordinates* to define the *design space*. A common example of such a space is the solid/liquid/gas phase space of a substance such as water (Figure 2.3).

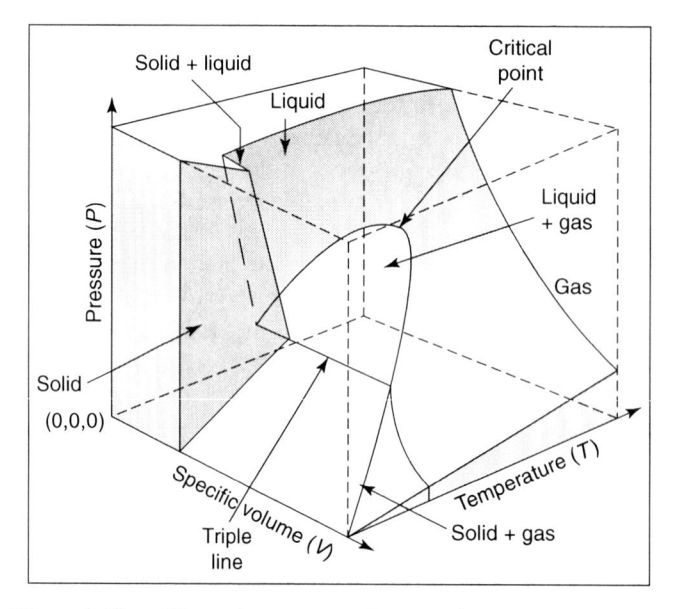

Figure 2.3. The solid/liquid/gas phase space diagram for water. http://commons.wikimedia.org/wiki/File:PvT_3D_plot_-_water.png.

As another example, we might have a design problem where we consider only elastic modulus E, tensile yield strength S_{yt}, and package volume V as the design variables or *design degrees of freedom* (DDOF). The design space would then look like Figure 2.4.

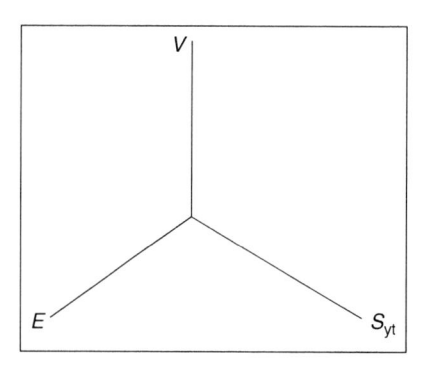

Figure 2.4. Abstract design space.

We would look for a solution inside this design space. Unfortunately, the solution space, and hence the number of possible solutions, is infinite. In order to converge on a single solution, we would need to bound (constrain) the solution space (just like in our particle problem) by placing requirements (limits) on one or more DDOF, for example, on volume, say $V < V_{max}$, some maximum volume (Figure 2.5).

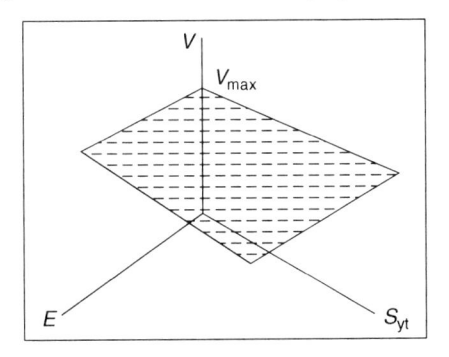

Figure 2.5. Design space constrained by limits on one of the DDOF.

Our search for a solution is now constrained to be below (bounded by) the plane defined by $V = V_{max}$. If we similarly constrain the other DDOF, you can imagine we would narrow our search to within a finite "box." The important point is that each of the DDOF can be manipulated within limits to narrow the solution space to achieve a solution, hopefully an optimal or near-optimal one. We talk of DDOF in a manner similar to that of CDOF. Of course, the number of DDOF could be quite large, which would no longer let us easily draw the space, but we can nonetheless still think in these terms.

According to Koen (1985), engineering design proceeds by the use of "heuristics." A heuristic is a usually nonanalytical procedure for achieving a solution (aka "rules of thumb"). By way of examples, several design rules are presented below that are generally applicable to all of engineering design:

- *KISS (Keep it simple, stupid)*: The idea here is that simplicity is good design, e.g., that simplicity will increase reliability while keeping costs low.
- *Make small changes in the state of the art*: In so many cases, especially where "getting it wrong" can lead to catastrophic failure and loss of life, product changes proceed very slowly and cautiously.
- *Leak before break*: Derived from hydraulic systems, this is a general principle that says: Design in such a way that you can easily find damage and flaws before they lead to catastrophic failure.
- *Fail safe*: Derived from structural design, this is a general principle that in practical application leads to adding redundancy, whether redundant load-carrying paths (components) or back-up computers and electronics.

2.2.2 Optimal Design

Design is always a trade-off among competing DDOF. We wish to maximize desirable DDOF, minimize undesirable DDOF, and limit or constrain the range of values of other

Chapter 2

DDOF. This is the *constrained optimization problem*. The design solution that best satisfies all of these competing requirements is the *optimal solution*. The optimal solution maximizes all desirable requirements while minimizing all undesirable outcomes (Johnson, 1980).

As a generic example, consider the following design problem: Maximize $Q(x)$ while limiting $C(x) \leq C_{max}$ for all values for $x \leq x_{max}$. Here Q is a desirable characteristic to be maximized, while C is a quantity to be limited. Graphically, the problem could be represented as in Figure 2.6.

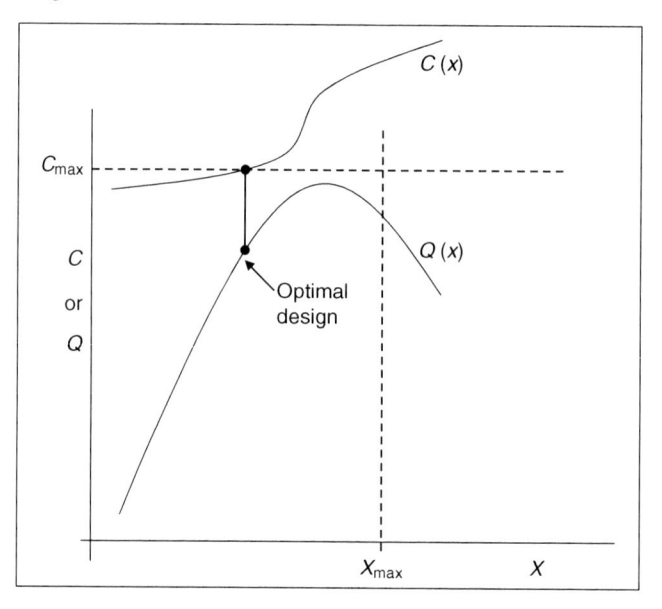

Figure 2.6. A generic constrained optimal design problem.

Example 2.1

Design a tray of minimum weight that can hold a volume V of a liquid.

Minimum weight corresponds to minimum mass for a given material. Let's consider a rectangular tray, perhaps stamped out of a thin metal sheet, as shown in Figure E.2.1.

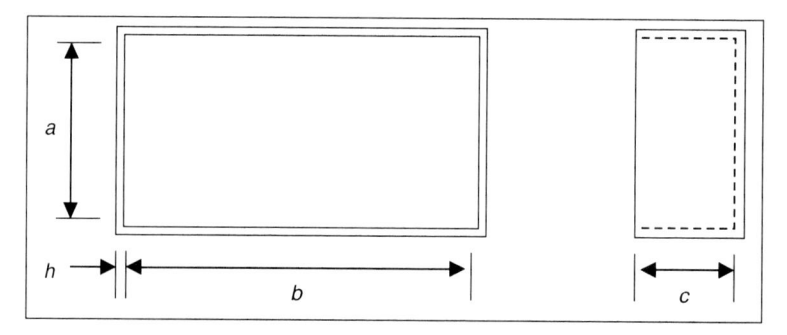

Figure E.2.1. A generic rectangular tray. Dimensions a, b, and c are internal dimensions.

Now $V = abc$, where a and b are the interior plan dimensions and c is the wall height. The tray thickness is h. (Note that the thickness h has no role in meeting the volume requirement.) The total amount of material mass is approximately

$$Q = (ab + 2bc + 2ac)h \tag{E.2.1}$$

Since a, b, and c are related through the original volume requirement V, we can eliminate one of the variables (say, a) such that (E.2.1) becomes

$$Q = \left(\frac{V_0}{c} + 2bc + 2\frac{V_0}{b}\right)h \tag{E.2.2}$$

A plot of Q vs. b would look like Figure E.2.2.

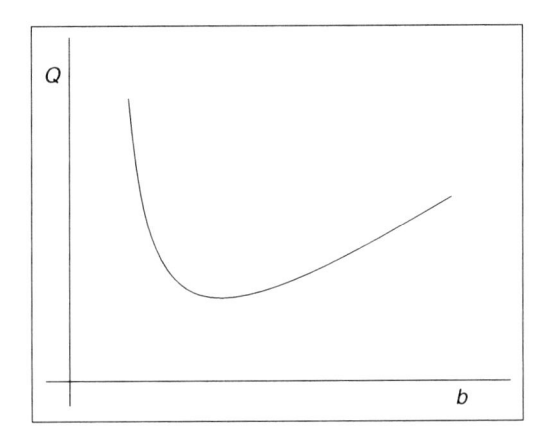

Figure E.2.2. Plot of Q vs. b.

The minimum or optimum value of b can be found from calculus as

$$\frac{\partial Q}{\partial b} = \left(2c - 2\frac{V_0}{b^2}\right)h = 2h\left(c - \frac{V_0}{b^2}\right) = 0 \Rightarrow b_{\text{opt}} = \sqrt{\frac{V_0}{c}} \tag{E.2.3}$$

Then from $V_0 = abc \Rightarrow a = V_0/bc$ and

$$a_{\text{opt}} = \frac{V_0}{b_{\text{opt}}c} = \frac{\sqrt{V_0^2}}{\sqrt{(V_0/c)}\sqrt{c^2}} = \sqrt{\frac{V_0}{c}} = b_{\text{opt}} \tag{E.2.4}$$

Then optimally $V_0 = a^2c$. This means that the optimum plan form that minimizes the mass for a given volume at fixed wall thickness and height is a square, $a = b$. (Optimizing against wall height c is left as an exercise for the student.) Figure E.2.3 shows the meaning of b_{opt}.

Chapter 2

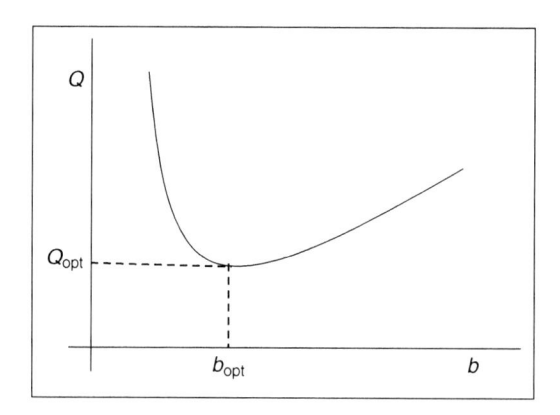

Figure E.2.3. The meaning of b_{opt}.

The material mass function Q then becomes

$$Q_{opt} = \left(\frac{V_0}{c} + 2\sqrt{\frac{V_0}{c}}\sqrt{c^2} + 2\frac{\sqrt{V_0^2}}{\sqrt{V_0/c}}\right)h$$

$$= \left(\frac{V_0}{c} + 2\sqrt{V_0 c} + 2\sqrt{V_0 c}\right)h \qquad (E.2.5)$$

$$= \left(\frac{V_0}{c} + 4\sqrt{V_0 c}\right)h = (a^2 + 4ac)h = a(1 + 4ac)h$$

Now h would probably be determined primarily for structural reasons, most likely stiffness, as well as manufacturability (e.g., die stamping).

Example 2.2

A steel rod of diameter d is used as a "link arm" in a piece of heavy equipment. The link length L can vary slightly such that $L_{min} \leq L \leq L_{max}$, and the weight must be less than w_{max}. The optimal design will maximize the link axial stiffness under these constraints.

The link weight and stiffness k are given by

$$w = \rho A L g \qquad (E.2.6a)$$

$$k = \frac{EA}{L} \qquad (E.2.6b)$$

where ρ is the mass density (kg/m^3), A is the cross-sectional area, g is the gravitational acceleration, and E is Young's modulus. A plot of w and k vs. L is given in Figure E.2.4.

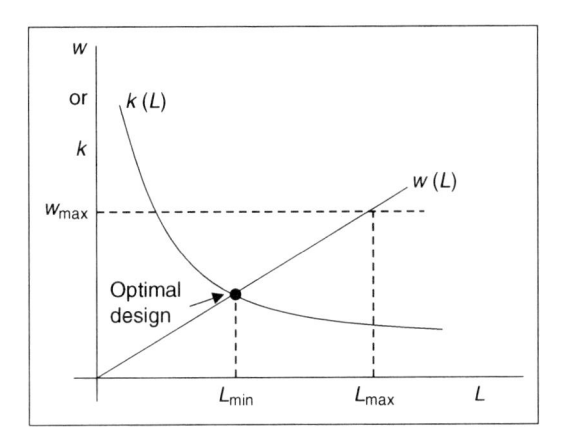

Figure E.2.4. Graph for Example 2.2.

Clearly, w_{max} at L_{max} will not provide the maximum k possible, but L_{min} satisfies the weight constraint while maximizing stiffness.

Reverse engineering is a type of inverse engineering design process. In reverse engineering, a physical artifact is studied to determine features and characteristics such as functionality, material composition, and manufacturing tolerances. Imagine a company has an old line of equipment for which the engineering drawings and other design information no longer exist, but the equipment still needs to be supported. The company may need to reverse engineer their own equipment to reconstruct the design information. More often, however, reverse engineering is used in market competition. Company A comes out with a new product and their competitor, Company B, studies the product through reverse engineering in order to deploy their own competitive product. In many industries, "there are no secrets"!

As a simple example, consider the following actual situation. A small "mom and pop" company (A) manufactures parts brushes that are used to clean automotive and other machine parts. Another company (B) wants to get into the parts brush business, so they obtain several of A's brushes, dissect them, send them out for chemical identification, analyze the brush bristles, etc. After some time, B's brushes hit the market and are virtually indistinguishable from A's (see Figure 2.7).

At first glance, B's brushes look nearly identical to A's. But closer examination reveals some differences. Notice that A's glue line is curved, while B's is straight across. Why is A's glue line curved? Is there a reason or was it just convenience? Does it matter that B's glue line is straight across? How many bristles are there in A's brush? Is there a reason for that exact number? If the number is slightly more or less in B's, does it matter? Are the brushes optimal, near-optimal, or far from optimal?

The point is that, as we stated earlier, the inverse problem is nonunique. It may be impossible to know from reverse engineering whether a certain feature (form) is that way because of functionality, or because it didn't matter, or because the engineer just missed it. These issues of nonuniqueness, form-function map, and optimality have important ramifications for BiE as we will see later.

In summary, the process of engineering design is iterative, one that by successive refinement leads to an (ideally) optimal solution to a set of design requirements within

Chapter 2

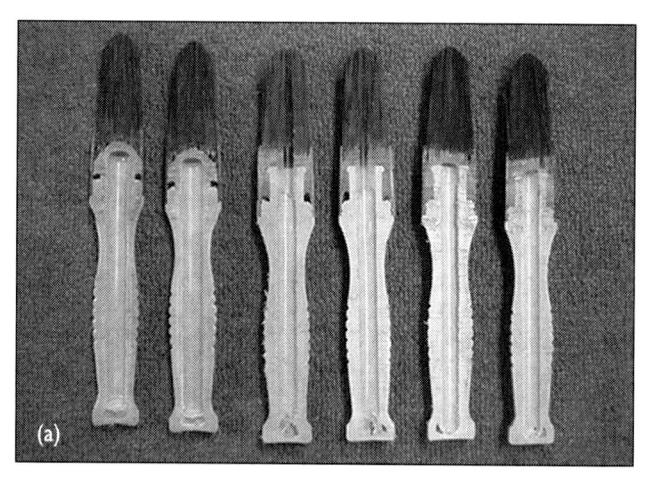

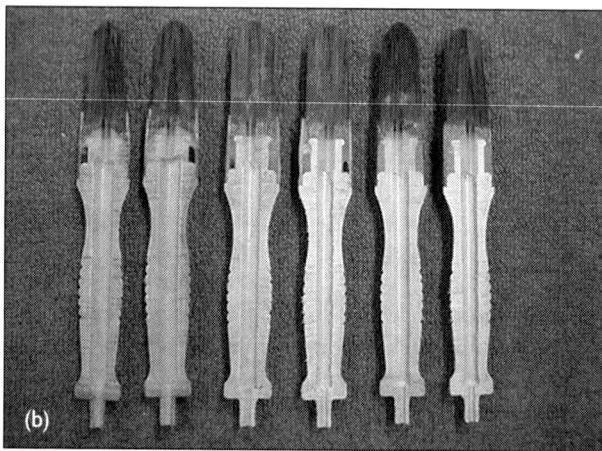

Figure 2.7. Sectioned halves of parts brushes: (a) parts brushes without flow channel and (b) parts brushes with flow channel. In (a) and (b), Company A's brush is the pair on the left, and Company B's brushes are the middle and rightmost pairs.

project constraints. Both forward and reverse design processes are used. (A reverse engineering design worksheet is included in Appendix 3; see also Problems at the end of this chapter.)

2.3 DESIGN IN NATURE

2.3.1 Design Process and Rules

As we mentioned at the beginning, the concept and practice of *design* are human constructs. The idea of *design in nature* will be very useful to us, but we must be cautious of applying this human process of invention and innovation too strictly to nature.

Recall that the engineering design process proceeds in stages but in an iterative fashion. The time scale of the iterations could be measured in weeks, months, or years. In nature, design is also an iterative process, but the time scale can be quite different: decades, centuries, and even millennia. This iterative process is part of a larger *evolutionary* process (Gregory, 2009; Mills, 2004).

Biological organisms reproduce and pass along genetic traits to their offspring. *Evolution* is the process of change in populations of offspring over generations. One of two principal mechanisms of change in biological organisms is *natural selection* (the other is *random genetic drift*). Beneficial traits that are better adapted to the living conditions become more common in descendant populations because the fitter individuals are likely to produce more offspring.

The beneficial traits that allow for and follow from adaptation can be thought of as nature's design rules, and these will form important components of the remainder of this text. A few rules that apply to structural design include the following:

- *Hierarchical design (HD)*: *Structural hierarchy* is a hallmark of biological systems. Its importance is severalfold. For one, structural hierarchy allows for variability in structural response at a much finer scale than would otherwise be possible. It allows for greater design freedom in layout, in reaching optimality. Optimizing a featureless, monolithic structure is difficult because the smallest practical module, which might be the structure itself, is rather large on the scale of stress variation.

- *Partition of structural function (PSF)*: The *PSF* is widely seen in natural systems demanding high performance, such as in animal flight. The wing of a bat is a good example (Figure 2.10). The aerodynamic loads on the wing are carried by a combination of tension and compression members. The wing covering is a flexible membrane-like skin that carries pressure loads essentially in tension. Compressive (and some flexural) loads are partitioned to the wing bones. In mammals, sinew and muscle carry tensile loads, while bones carry compression. This makes for a highly efficient structure that can approach a fully stressed, minimum-weight design. This is how modern sailboats are designed (sail and guy wires taking tensile loads, while the mast carrying compression and flexural loads). Cable-stayed bridges have been designed in a similar way for decades.

- *Low-modulus materials (LM)*: In many materials (at least in their elastic regime), whether natural or synthetic, stress and strain are related in such a way that if the strain increases so does the stress. However, the magnitude of the stress depends very much on the material's modulus or inherent stiffness. If the modulus is high, then even small increases in strain will result in high stress levels. Low-modulus materials can allow for large changes in strain while still keeping stress low. *Strength failures*, i.e., where the maximum stress in a structure exceeds a critical material strength, are less likely to occur in low-elastic-modulus materials. Low-elastic-modulus materials usually allow for greater stress redistribution around stress concentrations, thus again reducing strength failures. Spider webs are an excellent example in nature (Figure 2.8). As load-carrying elements become damaged or scissioned, the low modulus of the threads allows a major shape readjustment of the web; this allows for reduced stress in any of the remaining threads and continuation of prey capture functions (Alam and Jenkins, 2005). In human-engineered artifacts, pneumatic tires represent good examples. Because of their low elastic modulus, the tire deforms around an object, such as a rock, rather than trying to resist with stiffness and high stress. Low internal stress also reduces demands on the boundary supports. For equilibrium, boundary forces must balance the net internal forces (nominally stress times area). Again, the spider web is a good example. Webs may readily be attached to tree leaves or blades of grass due to the low internal stress.

- *Uniform stress (US)*: Stress concentrations are the boon of the structural engineer. Sharp discontinuities in geometry, say a notch or sharp corner, result in increased stress in the vicinity of the discontinuity, often by a factor of 3–10 or more. Failure is quite common as the stress concentration. Biological systems rarely have sharp corners. This results in a more uniform stress and less failure.
- *Just good enough (JGE)*: In engineering, minimalism is often a highly sought-after outcome. Minimizing redundancy such as in load paths, minimizing the number of components, and the like are perhaps indicative of striving for perfection. Nature, on the other hand, pursues solutions that are "just good enough." The redundancy evident in the load paths of spider webs (Figure 2.8) and mangrove trees (Figure 2.9) are excellent examples of just good enough design.

Although we will revisit nature's design rules in more detail throughout this book, several brief examples are provided in Figures 2.8–2.11 to help understand the concepts.

Figure 2.8. Spider web. Different varieties of threads have different capabilities and functions (PSF and HD). Redundant load paths (threads) provide damage tolerance (JGE). Flexibility absorbs impacts and shares load equally (LM). http://commons.wikimedia.org/wiki/File:Dewy_spider_web.jpg.

Figure 2.9. Mangrove roots. Roots play both structural and mass transport roles and are structured accordingly (HD). Redundant load paths (roots) allow for changing conditions (JGE). The roots live in an environment that would challenge many engineered structures.
http://commons.wikimedia.org/wiki/File:Mangrove.jpg.

Figure 2.10. Wings. Different elements support tensile loads, compressive loads, or control airflow (PSF). (a) Flexibility gives maneuverability and ability to change shape as needed (LM). http://en.wikipedia.org/wiki/File:Big-eared-townsend-fledermaus.jpg.
(b) Many feathers give damage tolerance and can be replaced individually (PSF).
http://commons.wikimedia.org/wiki/File:Green_Woodpecker_wing.jpg.

Figure 2.11. Palm trees. Leaves and trunk are flexible and help to handle wind loads (LM). Trunk is a multifunction structure (HD). http://commons.wikimedia.org/wiki/File:US_Navy_050709-N-0000B-004_Hurricane_Dennis_batters_palm_trees_and_floods_parts_of_Naval_Air_Station_%28NAS%29_Key_West%5Ersquo,s_Truman_Annex.jpg.

2.3.2 Evolution and Optimality

A central, crucial question for us is: Does evolution lead to optimality? Part of that answer depends on what we mean by optimality and our ability to recognize it. Is the entire biological system optimized or just one or more components of the system? Has evolution been at play long enough for the system to reach optimality? Is "near" optimality good enough? If biological solutions are not optimal, will they still help us expand the design space?

In biology the body of work that answers the central question in the affirmative is called *adaptationism*. Adaptationism is predicated on the assumption that all biological characteristics "are established in evolution by direct natural selection of the most adapted state, that is, the state that is in an optimum 'solution to a problem' posted by the environment" (Lewontin, 1983). Adaptationism and optimality are hotly debated in evolution theory today.

In a classic paper, Stephen Gould and Richard Lewontin (1979) use the example of the spandrels of St. Mark's Cathedral in Venice to question adaptationist thinking. Spandrels in this context are necessary by-products in fan-vaulted ceilings typical of gothic cathedrals such as St. Mark's. They result from the intersection of two rounded arches meeting at right angles. Elaborate artwork has been created on the spandrels. In "reverse engineering" of the cathedrals, would one be led to the conclusion that the spandrels were created for placement of elaborate artwork? In fact, they just simply exist and only the artwork makes them look like an adaptation.

Gould and Lewontin also give the example of chins. There are no genes for chins; rather, chins form as the result of two jaw bones (mandibles) growing together. Some chins are large, some small, some dimpled, etc. These are not adaptations, not optimizations—chins just exist.

Dennett (1996) argues that the spandrels could have been designed in other ways, all less aesthetically pleasing. Instead, they are "minimum surfaces" (as in soap film stretched across a wire frame) and as such optimal from the perspective of material cost. Their smooth surface is also optimal for supporting the religious art so important to the cathedral. Hence, they are adaptations, chosen from a set of design alternatives.

Because of evolutionary history, natural designs never start with a clean slate. In engineering, however, we often take advantage of an "ah-ha" momentum and develop a "revolutionary" technology. Natural designs are also highly constrained: it's like asking an engineer to design an automobile that starts off tiny and becomes full size but has to work at every step along the way (Full, 2002).

> Just copying nature directly is absolutely not what you want to do . . . instead you need to be inspired by biology (Full, 2002).

KEY POINTS TO REMEMBER

- Engineering design is a creative human activity that can proceed in a forward or reverse manner.
- Engineering design proceeds by heuristics or rules of thumb.
- Design is a trade-off among competing design degrees of freedom, and ideally an optimum trade-off.
- We can think of nature as a designer and distill design rules accordingly.
- BiE is predicated on natural solutions being optimal, but the reality is a bit more complicated.
- The central reason for the study of BiE is to expand the design space of possible engineering solutions.

REFERENCES

Alam, M. S., and C. H. Jenkins. 2005. "Damage Tolerance in Naturally Compliant Structures." *J. Damage Mech.* 14:365–384. doi: 10.1177/1056789505054313; http://ijd.sagepub.com/content/14/4/365.

Dennett, D. C. 1996. *Darwin's Dangerous Idea*. Simon & Schuster.

Dieter, G. E. 2000. *Engineering Design*, 3rd ed., chap. 8. McGraw-Hill.

Dym, C. L., and P. Little. 2004. *Engineering Design*. Wiley.

French, M. J. 1988. *Invention and Design: Design in Nature and Engineering*. Cambridge University Press.

Full, R. 2002. TED (http://www.ted.com/talks/robert_full_on_engineering_and_evolution.html).

Chapter 2

Gould, S. J., and R. C. Lewontin. 1979. "The Spandrels of San Marco and the Panglossian Paradigm: A Critique of the Adaptationist Programme." *Proc. Roy. Soc. Lond. B* 205:581–598.

Gregory, T. R. 2009. "Understanding Natural Selection: Essential Concepts and Common Misconceptions." *Evol. Educ. Outreach* 2 (2):156–175. doi: 10.1007/s12052-009-0128-1; http://www.springerlink.com/content/2331741806807x22/.

Jenkins, C. H., and S. Khanna. 2005. *Mechanics of Materials: A Modern Integration of Mechanics and Materials*. Elsevier.

Johnson, R. C. 1980. *Optimum Design of Mechanical Elements*, 2nd ed. Wiley.

Koen, B. V. 1985. *Definition of the Engineering Method*. ASEE.

Lewontin, R. C. 1983. "Elementary Errors About Evolution. Review of [Dennett 1983]." *Behav. Brain Sci.* 6:367. doi: 10.1017/S0140525X00016538; http://journals.cambridge.org/action/displayAbstract?fromPage=online&aid=6719716.

Mills, C. L. 2004. *The Theory of Evolution*. Wiley.

Shigley, J. E., and C. R. Mischke. 2001. *Mechanical Engineering Design*, 6th ed. McGraw-Hill.

Ullman, D. G. 1997. *The Mechanical Design Process*. McGraw-Hill.

PROBLEMS

2.1. A mass particle is constrained to a circular trajectory of radius R in x–z space. The center of the circle resides at the location (x_0, z_0), not $(0, 0)$. Show that the distance of the particle from the origin of coordinates is given by

$$|r| = \sqrt{x_0^2 + z_0^2 + 2R(x_0 \cos \theta + z_0 \sin \theta) + R^2}$$

2.2. A compound pendulum consists of two rigid massless links of lengths L_1 and L_2 (hinged together) and one massive bob of mass m (Figure P.2.2). How many coordinate degrees of freedom are involved in this configuration if the motion is restricted to x–y space?

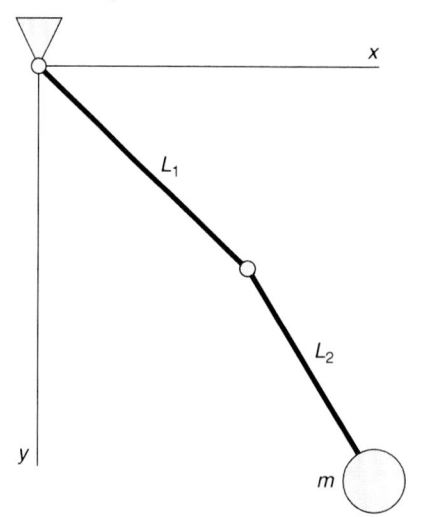

Figure P.2.2

2.3. A dime rolls on a tabletop without slipping. How many coordinate degrees of freedom are required to completely locate a position on the face of the coin, say the tip of Roosevelt's nose?

2.4. The *spherical pendulum* consists of a mass m connected to a swivel through a massless rigid link l (see Figure P.2.4). The spherical coordinates θ and ϕ measure the rotation from and about the z-axis, respectively. The Equations of Motion in matrix form are as shown.

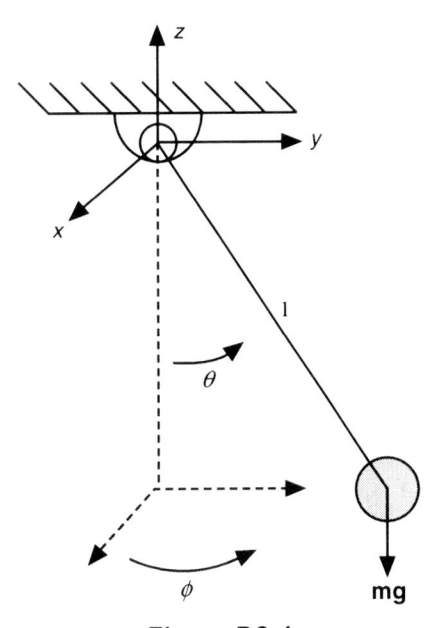

Figure P.2.4

$$\begin{bmatrix} ml^2 & 0 \\ 0 & ml^2 \sin\theta \end{bmatrix} \begin{bmatrix} \ddot{\theta} \\ \ddot{\phi} \end{bmatrix} + \begin{bmatrix} ml^2 \sin\theta \cos\theta\, \dot{\phi}^2 \\ 2\, ml^2 \sin\theta \cos\theta\, \dot{\theta}\dot{\phi} \end{bmatrix} + \begin{bmatrix} mgl \sin\theta \\ 0 \end{bmatrix} = 0$$

How many CDOF are required in this case? If the angle ϕ is constrained to be constant, that is $\phi = \phi_0$, show that the equations of motion for the spherical pendulum degenerate to the equation of simple pendulum. What is the physical meaning of ϕ = constant?

2.5. Discuss the "cradle-to-grave" considerations for an automobile. Give examples of DDOF.

2.6. Discuss the following engineering rules of thumb (after Koen):

   a. Use feedback to stabilize an engineering design.

   b. Allocate sufficient resources to the weak link.

   c. At some point in the project, freeze the design.

   d. The yield strength of a material is the 0.02% offset on the stress–strain curve.

2.7. Give at least two examples of each technique, one from biology and one from engineering: HD, PSF, LM, JGE, and US.

2.8. Rework Example 2.1, this time optimizing against the wall height c.

Chapter 2

2.9. Reconsider Example 2.1 but with a circular plan form.

2.10. Design a rectangular tube of length L, wall thickness h, and cross section $a \times b$, which has minimum mass for a given interior volume. Both L and h are finite and nonzero. The dimensions a and b are outside dimensions.

2.11. Design a cylindrical tube of length L, wall thickness h, and outside diameter d, which has minimum mass for a given interior volume. Both L and h are finite and nonzero.

2.12. A steel rod of diameter d is used as a "link arm" in a piece of heavy equipment. The link area A can vary slightly such that $A_{min} \leq A \leq A_{max}$ and the weight must be less than w_{max}. The optimal design will minimize the link axial stress (P/A) under these constraints.

2.13. *Group design activity.* Choose a product with moderate complexity (e.g., office product, construction tool, sporting good) and reverse engineer the product following the Reverse Engineering Worksheet (Appendix 3).

CHAPTER 3

Structures and Materials in Engineering

> You can make a bad structure from good materials, but you can't make a good structure from bad materials.
>
> —*Anonymous*

A particular application of bio-inspired design is to expand the structural design space by learning from natural structures. Structures in nature have much to teach us; even simple structures like trees can adapt to their environments, grow, and self-heal. Nature is an outstanding integrator of structure and materials. In order to prepare us for learning from nature, in this chapter we review basic concepts in structures and materials in engineering, such as geometry, boundary conditions, methods of structural analysis, and material constitutive relations.

3.1 OVERVIEW OF STRUCTURES

3.1.1 Structural Loads

In structural design, we use the term "loads" to mean forces and moments either applied to the structure <u>externally</u> on the boundaries (*surface loads*) or developed <u>within</u> the structure (*body loads*).

Loads are further considered to be either static or dynamic. Static loads are loads that do not depend on time, i.e., they are of constant magnitude, direction, and location. Although it might seem that certain structures are static, e.g., on a civil structure such as a building, this is rarely the case. "Live loads" from occupant activity, wind loads, seismic (earthquake) loads, thermal cycling, etc., all may give rise to a dynamic load environment. However, if the loads vary slowly with time, they are often considered *quasi-static*, and taken as static loads. (*Slowly* usually means that structure inertia force due to accelerations may be neglected with respect to the difference of the externally applied forces and the internal resistance forces.)

Dynamic loads are divided into two main categories:

1. *Steady-state* loads are loads that maintain the same character (frequency, amplitude, etc.) over the long term.
2. *Transient* loads are loads that change their character (e.g., they may *decay*) with time.

Common structural loads are summarized in Table 3.1.

Table 3.1. Summary of common structural loads

Surface loads (Common name)	Units SI (US)	Body loads	Units SI (US)
Concentrated force ("Point load")[a]	N (lb)	Gravitational force ("Gravity load" or "weight")	N (lb)
Distributed force ("Line load")[a] ("Pressure")	N/m (lb/in.) N/m^2 (psi)	Thermal stress ("Thermal load")	N/m^2 (psi)
Moment or couple ("Bending or torsion moment")	N m (lb in.)		

[a] In principal, "point" and "lines" loads cannot exist, since any load will act over a finite spatial "area," no matter how small; however, if that area is small relative to the structure size, then for practical purposes the load may be considered a "point" load or a "line" load.

The *orientation* of the load on a structural member is also important. Although this issue will be discussed in more detail in later chapters, a brief summary is given in Figure 3.1.

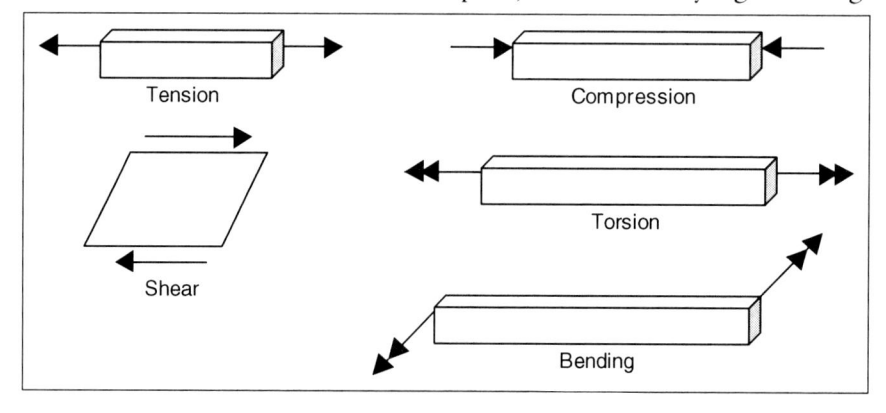

Figure 3.1. Simple load orientations. Single-headed arrows are forces; double-headed arrows are moments (that follow the right-hand rule).

3.1.2 Structural Form

Carrying loads is the primary *function* of a structure, and it is this characteristic that largely determines the *form* of the structural elements. Loads carried include tensile and compressive axial loads, shear loads, torsion moment loads, bending moment loads, distributed loads, gravity loads, and thermal loads.

Structural forms fall into two major categories: *line*-forming and *surface*-forming; surface-forming elements may be further subdivided into *area*-forming and *volume*-forming elements (see, e.g., Schodek, 2001). A taxonomy of structural elements is given in Table 3.2.

Table 3.2. A taxonomy of structural elements

FUNCTION	FORM						
	Line-forming			Surface-forming			
				Area-forming		Volume-forming	
Loads carried:	Cable	Rod	Beam	Plane membrane	Plate	Curved membrane	Shell
Tensile axial	✓	✓	✓	✓	✓	✓	✓
Compressive axial	—	✓	✓	—	✓	—	✓
Direct shear	—	—	✓	—	✓	—	✓
Torsion moment	—	✓	✓	—	✓	—	✓
Bending moment	—	—	✓	—	✓	—	✓
Distributed force	✓		✓	✓	✓	✓	✓
Thermal	✓	✓	✓	✓	✓	✓	✓

A review of Table 3.2 reveals that beams, plates, and shells are structural elements fully capable of carrying all types of loads. Specialty elements such as cables, rods, and membranes carry limited types of loads. (*Note*: *finite element analysis* codes used in structural analysis have libraries of structural elements based upon a similar taxonomy.)

Line-forming elements (LFEs) are slender structures having one spatial dimension (length) significantly greater than any other dimension (width, height, thickness, etc.). Schematically, for analysis purposes, LFEs may be represented as lines, either straight or curved as required. LFEs may carry loads in tension, compression, torsion, bending, or some combination of these, depending on the nature of the applied loads, structural geometry, material properties, and boundary conditions. LFEs may form *axial*, *torsional*, or *bending* structures such as *rods*, *cables*, and *beams*.

1. *Rods* carry concentrated axial forces in either tension or compression (no moments or transverse forces). Thus, they are *axial structures*. Only *pinned* ("simply

supported") boundary conditions are required. Cross-sectional geometries may take any shape, but simple shapes are most common, such as circular and rectangular. The cross section may be solid or hollow.

2. *Cables* carry concentrated axial tensile forces (but not compressive axial forces—"you cannot push a rope"!), as well as concentrated or distributed transverse forces (but not moments!). In the latter case, cables deform under the action of these transverse loads in such a way that they remain *axial* structures (specifically "no-compression axial structures"). (Of course, in the former case of axial loads, cables are by definition axial structures.) As with rods, only pinned boundary conditions are required. Cross-sectional geometries may be of any shape, but solid circular cross sections are common.

3. *Beams* are the most complete LFEs because they can carry axial compressive or tensile forces as in rods or cables, as well as transverse concentrated or distributed forces as in cables; moreover, they can carry torsion and bending moments. In cases where moments are applied, at least one boundary condition must be able to support moment reactions (e.g., a "fixed" or "built-in" condition). Cross-sectional shapes may be of any geometry, solid or hollow.

Surface-forming elements (SFEs) are thin structures having two spatial dimensions (length and width) significantly greater than the third dimension (thickness). Schematically, for analysis purposes, SFEs may be represented as surfaces, either straight or curved as required. SFEs may carry loads in tension, compression, torsion, bending, or some combination of these, depending on the nature of the applied loads, structural geometry, material properties, and boundary conditions. SFEs may form axial, torsional, or bending structures such as *plates* and *shells*. Being more complicated structures, SFEs are outside the scope of this book, and will not be considered further.

3.1.3 Boundary Conditions

It should be obvious that if a structure is not tied down somewhere, it will not be able to carry loads in many situations. Imagine trying to hang a weight from a hook not connected to anything! Thus, we see that the conditions at the structure boundaries, i.e., the *boundary conditions*, are critically important in structural design and analysis.

A number of "idealized" boundary conditions can be defined. (We say "idealized" because boundary conditions on real structures rarely meet these definitions exactly.) In general, boundary conditions resist either *translation* or *rotation*, or both, in one or more directions. In order to do this, they must generate resisting forces (translation) or moments (rotations). In other words, translation and/or rotation are *resisted* at the boundary by the generation of either a force and/or moment, respectively. (Mathematically, boundary conditions provide additional *equations of constraint*.)

If there are $\begin{Bmatrix} \text{less} \\ \text{more} \end{Bmatrix}$ constraints than the minimum required, the structure is $\begin{Bmatrix} \text{underconstrained} \\ \text{overconstrained} \end{Bmatrix}$, i.e., the structure is *statically indeterminate* in that there are $\begin{Bmatrix} \text{more} \\ \text{less} \end{Bmatrix}$ independent equations available than unknowns to be determined.

A number of common boundary conditions are provided in Table 3.3.

Table 3.3. Common boundary conditions

Name (other names)	Translation constraint	Rotation constraint
Free—2D or 3D	None	None
Fixed (clamped)—2D or 3D	No translation ($u = v = w = 0$)	No rotation ($u' = v' = w' = 0$)
Ball and socket (swivel)—3D	No translation ($u = v = w = 0$)	None
Simple (pinned)—2D	No translation ($u = v = 0$)	None
Simple (roller)—2D	No translation $\perp$ to roller surface (e.g., $v = 0$)	None

Note: The prime or ' symbol indicates differentiation with respect to a spatial coordinate, i.e., a *gradient* or "slope." u, v, and w are displacements in the x, y, and z directions, respectively.

3.1.4 Methods of Structural Analysis

When we think of a structure carrying loads, two primary questions must be asked:

- Is it *strong* enough?
- Is it *stiff* enough?

In other words, we wish to know how well the primary functional requirement of a given structure—carrying loads—is performed; this is the primary role of *structural analysis*. (A number of other questions must be asked as well, such as: Is it light enough? Cheap enough? Repairable enough? …)

Strength and stiffness are independent quantities (they are *design degrees of freedom*, a concept discussed in Chapter 2) that depend on the constituents of the material in different ways. We'll come to understand these concepts more in later sections. At this point, we just mention that strength considerations are usually of greater importance in structures than stiffness. If you carry a load with a large rubber band, you are usually not too worried if it stretches quite a bit, as long as it doesn't break (i.e., as long as it is strong enough). However, if the band stretches too much, your arms may not be long enough to keep the load from interfering with the ground, and now stiffness (or lack of it!) does become of concern.

An important issue we cannot avoid is: "How much is enough?" Unfortunately, the answer is a somewhat unsatisfying—"It depends." It depends on the application, the failure mode, how catastrophic would be the failure (e.g., loss of human life), the past history of similar designs, how well known are the loads and material properties, etc. In any case, we must always compare what we have (stress, strain, deflection—the *load effects*) to what we can allow (strength, stiffness—*the resistance*).

Structural analysis consists of four fundamental parts:

1. *Configuration.* Geometry, loads, and boundary conditions are essential elements that must be considered early in the analysis.
2. *Equilibrium.* Here we consider forces, moments, and the application of Newtonian mechanics (or the potential and kinetic energies and the application of Lagrangian mechanics), and stress. If the body is in *global* equilibrium, then every *local* particle of the body is also in equilibrium. Equilibrium implies negligible accelerations (inertia forces), and hence implies *static analysis.* (*Dynamic analysis* would consider the motion of the body, requiring inertia forces in the full *equations of motion.*)
3. *Deformation.* Here we consider the geometry of material particle displacement and the concept of strain. We assume that any material is *continuous* and fully populated with *material particles* (the *continuum hypothesis*). We further assume in this chapter that deformations are small enough that only a linear analysis is required.
4. *Constitution.* Stress and strain are *dual* quantities that are intimately related within a given material/structural system. Parts (1) to (3) are independent of specific material behavior, which is necessarily supplied through the *constitutive relations* (*stress–strain relations*).

3.2 APPLICATION TO ENGINEERING STRUCTURES

3.2.1 Configuration

In Figure 3.2, an axial structure called a *bar, link, rod,* or *strut* of round cross-sectional shape is shown. A force P is applied to one end of the rod, resulting in a displacement u. The other end of the rod has a boundary condition described as either fixed or built-in. The rod has geometric properties of length L and constant cross-sectional area A; material properties are *mass density* ρ, *elastic modulus* E_t, and *tensile strength* S_t.

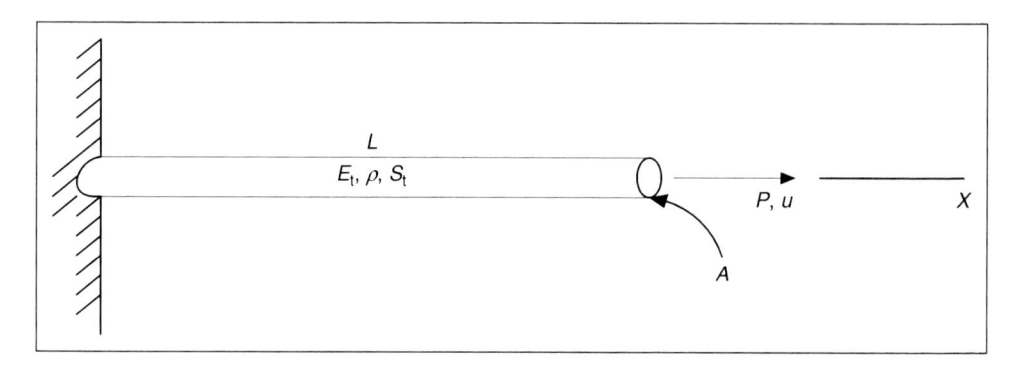

Figure 3.2. Definition sketch of an axial structure.

(Keep in mind that although we have shown a rod with *constant, circular* cross section, primarily for convenience and simplicity, a rod may have a *nonconstant* and/or *noncircular* cross section. However, to remain an axial structure, the axis of every cross section, which is the line passing perpendicularly through the centroid of that cross section, must remain colinear. Structures with constant cross-sectional area, or more generally constant area moment of inertia, are called *prismatic* structures. Also, keep in mind that quantities such as force and displacement are *vector* quantities but in the discussion to follow there is no confusion by representing them by their scalar magnitude only.)

As another example, consider a prismatic torsion structure of circular cross section with radius R (Figure 3.3). The torsion structure is straight and relatively narrow, i.e., the dimension of any cross section is small compared to the element length L. The cross section is characterized by its area A and *polar moment of inertia (polar second moment of area) J*.

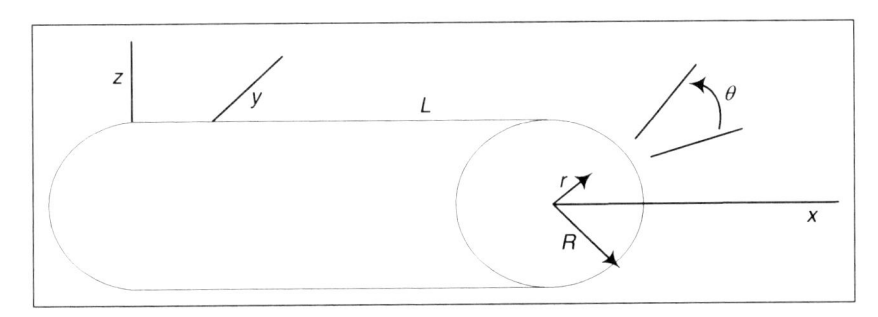

Figure 3.3. Torsion structure of constant circular cross section. Any point on the circular cross section $x =$ constant can be identified by the cylindrical–polar coordinates (r, θ, x).

Loads

The loads on an axial structure must be *coaxial* loads (proved below), i.e., loads acting coaxially with the significant structural axis (x axis here), such as the *concentrated force* applied to the ends of the rods above. The force shown in Figure 3.2 is *tensile*, since it acts to create tension in the bar, i.e., acts to displace material particles further away from one another. The force can also be *compressive*, so as to create a compressive action on material particles, i.e., acting to displace material particles closer together than in their unloaded configuration. No moments are applied to axial structures.

The torsion structure is loaded by a *twisting moment* or *torque* M. In practice, this could arise from a moment of magnitude Pd applied to the shaft as shown in Figure 3.4.

Boundary Conditions

In the case of tensile axial structures, we readily allow the case of pinned-free boundary conditions. In the compressive case, pinned–roller boundary conditions are common. (There are issues of stability to be addressed, but these are outside the scope of this chapter; see Jenkins and Khanna, 2005.) Common boundary conditions for torsion structures are pinned–pinned and fixed (clamped)–free; these are also common boundary conditions for beams.

33

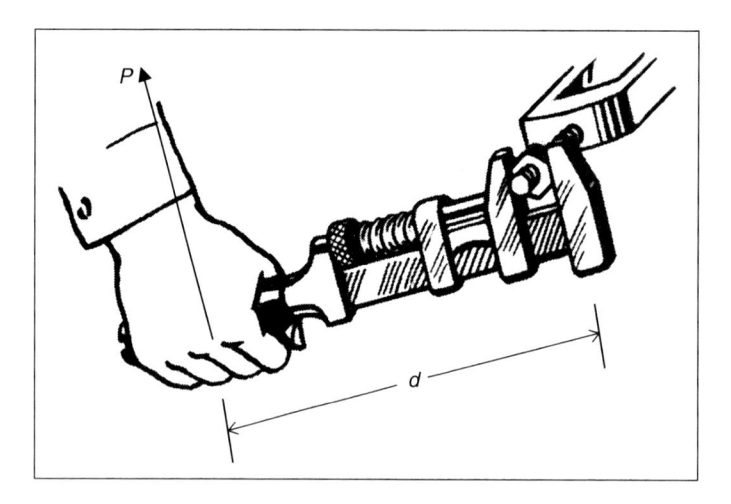

Figure 3.4. A moment being applied to a shaft by a force P on a wrench of length d. http://commons.wikimedia.org/wiki/File:Monkey_Wrench_%28PSF%29.png

3.2.2 Equilibrium

We now consider a *free-body diagram* (FBD) of a rod (which should include the externally applied forces and moments, the internal reaction forces and moments, and the necessary geometry and coordinate descriptions necessary to apply Newton's laws). The right end of the rod is "freed" from the body at some position x from the origin and the corresponding FBD looks as shown in Figure 3.5.

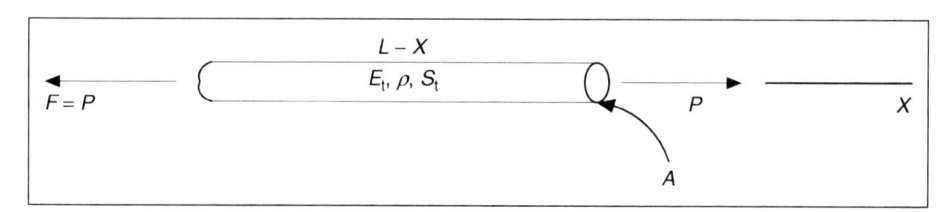

Figure 3.5. Global free-body diagram of an axial structure.

Shown on the FBD is the internal reaction force F, which from global static equilibrium analysis

$$\sum F = 0$$

can be shown to be

$$P - F = 0 \Rightarrow F = P$$

Furthermore, global moment equilibrium

$$\sum M = 0$$

shows that *F* and *P* are *colinear* (*coaxial*), since they must present no net *couple* (moment) on the rod.

We now look more closely at the stress in an axial structure and the corresponding local equilibrium. Consider the local FBD or *stress element* carved out of the cross section *A*, shown in Figure 3.6.

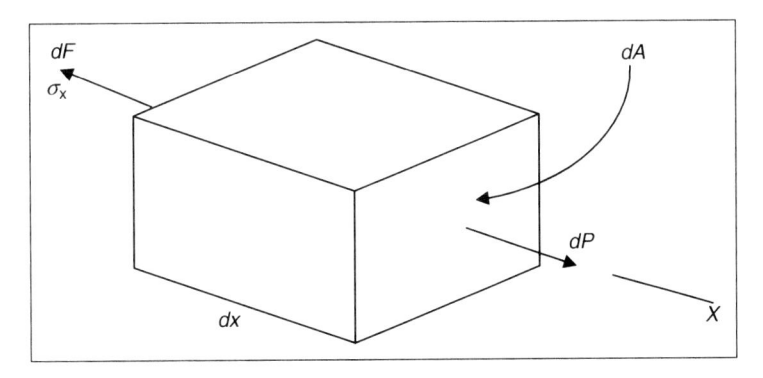

Figure 3.6. Local free-body diagram of an axial structural element.

Now an increment of internal force, *dF*, can be seen to act over an increment of area, *dA*, which balances the increment of applied load *dP* (i.e., *dF* = *dP*). We take as a fundamental postulate (known formally as *Cauchy's stress principle*) that the externally applied load is resisted internally (to enforce local equilibrium) by a stress σ_x such that

$$dF = \sigma_x dA \tag{3.1}$$

Based on the assumptions and discussion for an axial structure given above, the stress σ_x is assumed to be constant as long as no intermediate axial loads are applied, hence a simple integration gives

$$\sigma_x = \frac{F}{A} \tag{3.2}$$

(The *stress state* in an axial structure is considered to be relatively simple, in that it is assumed that only *normal* stresses exist on cross sections normal to the rod axis. The *stress distribution* is also considered to be relatively simple, except in a local region around where the load is applied, which we will ignore—this is called *St. Venant's principle*. Keep in mind that what we really calculate with equation (3.2) is an *average* stress acting *everywhere* over the cross section of the bar, even though stress is fundamentally a measure of the load response by a structure *at a point*. However, as mentioned above, St. Venant's principle allows us to disregard this detail in many cases, and we use that convenience here.)

It is possible that on the opposite face of the stress element, the incremental force is not *dF* but has changed by some *gradient* (spatial rate of change) of the force, *dF/dx*, that is

$$dF + \left(\frac{dF}{dx}\right)dx$$

Chapter 3

due to a local change of loading across the span *dx* of the element. Then a local application of equilibrium, i.e., summation of (incremental) forces, in the *x* direction gives

$$dF - \left[dF + \left(\frac{dF}{dx}\right)dx\right] = 0 \Rightarrow \frac{dF}{dx} = 0$$

since $dx \neq 0$. Now, substituting $\sigma_x \, dA$ for dF from equation (3.1) gives

$$\frac{d(\sigma_x dA)}{dx} = 0$$

or, since $dA \neq 0$,

$$\frac{d\sigma_x}{dx} = 0 \tag{3.3}$$

This is a very fundamental and general requirement for local equilibrium, i.e., the <u>gradient of the stress equals zero</u>.

Local and global equilibrium can be related through equations (3.4):

$$F = \iint_A dF = \iint_A \sigma_x dA = P$$

$$M_y = \iint_A z \cdot dF = \iint_A \sigma_x z \, dA = 0 \tag{3.4}$$

$$M_z = -\iint_A y \cdot dF = -\iint_A \sigma_x y \, dA = 0$$

In the torsion case, the applied couple is resisted internally by a shear force magnitude *F*. Figure 3.7 shows a cylindrical differential element of the shaft of length Δx and

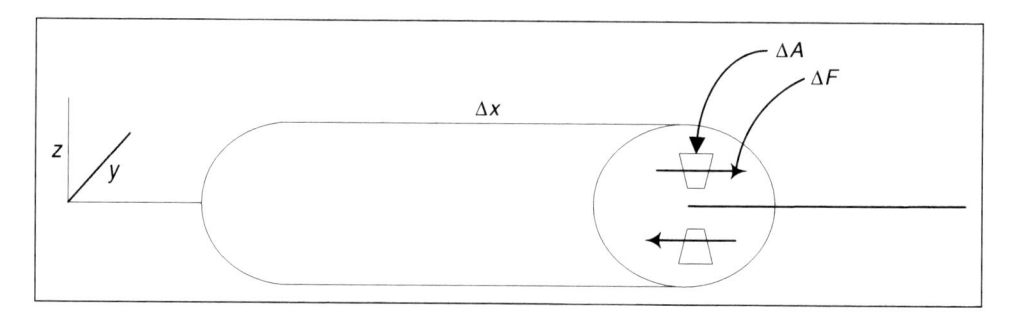

Figure 3.7. Internal force system on a differential shaft element consisting of a shear force ΔF.

cross section ΔA (radius $r < R$), which is coaxial with the original shaft of radius R. The associated direct shear stress on the element face is $\tau_{x\theta}$ given by

$$\tau_{x\theta} = \lim_{\Delta A \to 0} \frac{\Delta F}{\Delta A} = \frac{dF}{dA} \quad (3.5)$$

where ΔF is the incremental shear force. Due to symmetry, F or $\tau_{x\theta}$ can only be functions of r and x, but not θ.

Local–global equilibria are related by

$$M = \int_A r\,dF = \int_A r\tau_{x\theta}\,dA \quad (3.6)$$

Example 3.1

A solid circular cylinder of length L and cross-sectional area A needs to be pulled from the ground within which it is completely embedded as shown in Figure E.3.1a. The ground exerts a distributed axial force on the cylinder of the form $p(x) = p_0(x/L)$. Determine the stress in the rod as a function of position along the rod length and the pull force P. (This could be a simple model of a plant root!)

1. Configuration—the configuration is shown in Figure E.3.1.
2. Global equilibrium:

$$\sum F_x = 0 : P = \int_0^L p(x)\,dx = \frac{p_0}{L}\frac{L^2}{2} = \frac{p_0 L}{2} \Rightarrow p_0 = \frac{2P}{L}$$

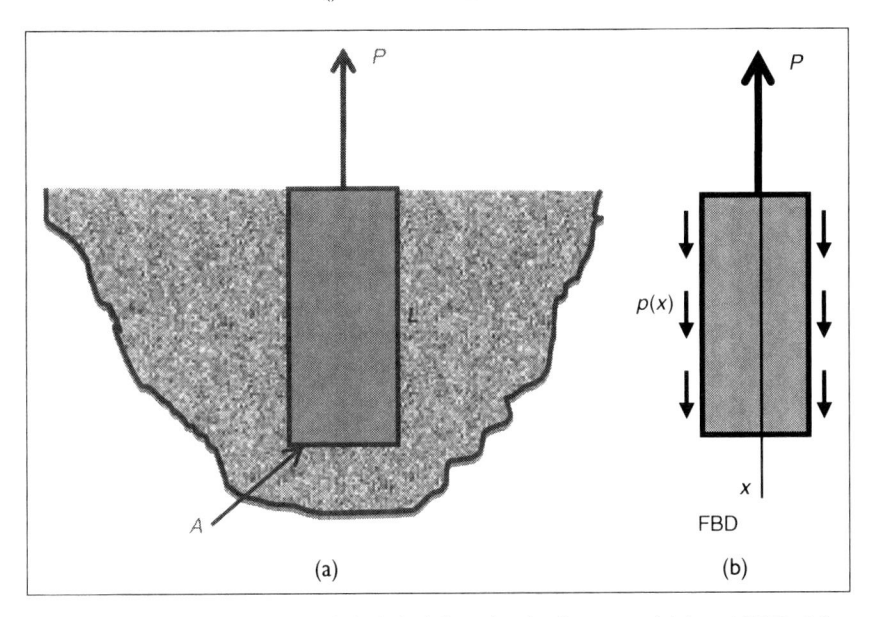

Figure E3.1. (a) Configuration; (b) global free-body diagram; (c) local FBD; (d) stress variation along the structure length.

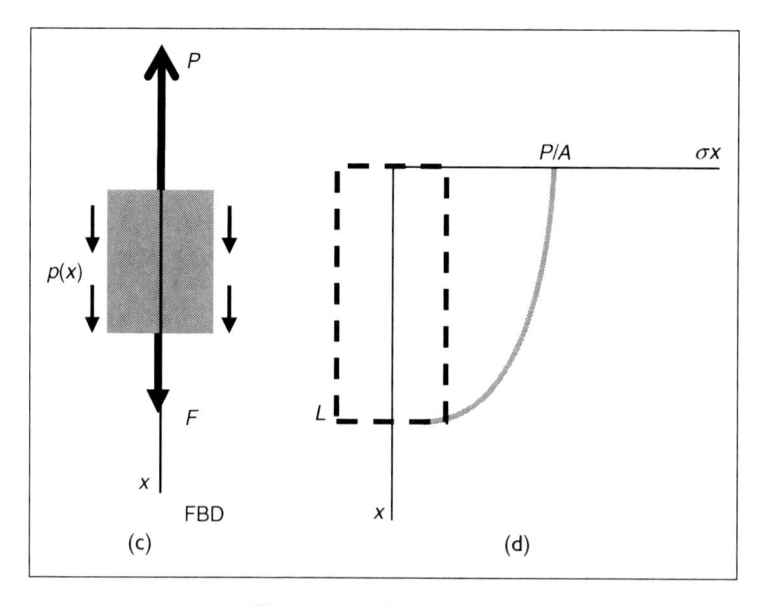

Figure E3.1. *Continued*

then
$$p(x) = p_0 \frac{x}{L} = \frac{2P}{L^2} x$$

3. Local equilibrium:
$$\sum F_x = 0: \quad -P + \int_0^x p(x)\,dx + F(x) = 0$$
$$F(x) = P - \int_0^x \frac{2P}{L^2} x\,dx = P - \frac{2P}{L^2}\frac{x^2}{2} = P\left(1 - \frac{x^2}{L^2}\right)$$
$$\sigma(x) = \frac{F(x)}{A} = \frac{P}{A}\left(1 - \frac{x^2}{L^2}\right)$$

3.2.3 Deformation

Under the action of the tensile force F, a material particle located at position x in the rod will displace an amount $u(x)$ in the $+x$ direction. The corresponding *strain* is a strictly geometrical quantity. The simplest definition we can take is that strain (at a point!) is the comparison of a local change in length between two material particles relative to some "gage length," which is taken as the length between the same particles in the undeformed state, namely dx. If we take the deformed length to be dx^*, then the strain in the x direction is defined as

$$\varepsilon_x \equiv \frac{dx^* - dx}{dx} \tag{3.7a}$$

$$\varepsilon_x = \frac{du}{dx} \tag{3.7b}$$

(There is no single, unique definition for strain. We use here the common definition of one-dimensional (1D) *engineering strain*.) This is a very fundamental and general requirement of deformation that relates strain to displacement, and is called the *strain–displacement relation* (although it is more correctly the strain–displacement <u>gradient</u> relation). These are also called the *kinematic relations*.

We see that the strain ε_x is a normal (here tensile or extensional) strain in the axial direction. The axial displacement <u>anywhere</u> along the rod (at any cross section) can then be found from

$$u(x) = \int_0^x \varepsilon_x \, dx \tag{3.8}$$

with the total increase in the rod length given by

$$u(L) = \int_0^L \varepsilon_x \, dx \tag{3.9}$$

We emphasize that for a rod modeled as a 1D structure, all fibers (which are axial fibers) deform identically. (Strains also can be generated by unconstrained thermal expansion or contraction; see, e.g., Boley and Weiner, 1997.)

We assume the pure torsion or twisting deformation to be smooth and regular. For a torsion element undergoing <u>small</u> twisting deformations, it is reasonable to further assume the following:

1. Cross sections originally plane and normal to the element long axis remain plane and normal to the axis after deformation.
2. The originally symmetric cross sections remain symmetric after deformation.
3. No displacement occurs in either the axial or radial directions.

For the circular shaft, these assumptions would imply that:

1. A diameter $2R$ before deformation remains a diameter after deformation with the same value $2R$, i.e., $\varepsilon_r = 0$.
2. All diameters at any given section $A(x)$ rotate through the same *twist angle* $\phi(x)$; hence, $\gamma_{r\theta} = \varepsilon_\theta = 0$.
3. The length of the shaft remains unchanged, hence $\varepsilon_x = \gamma_{rx} = 0$.

Thus, if a given cross section $A(x)$ rotates by an angle $\phi(x)$, a nearby cross section $A(x + \Delta x)$ rotates by an angle $\phi(x + \Delta x) = \phi(x) + \Delta\phi$ (and $A(x - \Delta x)$ rotates by an angle $\phi(x - \Delta x) = \phi(x) - \Delta\phi$). Also, the only nonzero strain is $\gamma_{x\theta}$.

To understand the physical meaning of the shear strain $\gamma_{x\theta}$, consider the cylindrical differential element of radius r ($<R$) and length Δx previously introduced in Figure 3.7 (shown again now in Figure 3.8).

Chapter 3

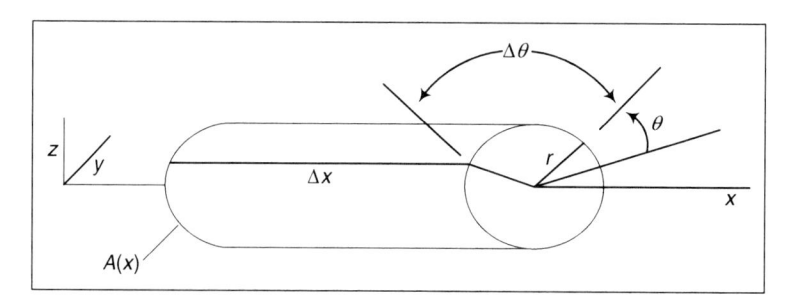

Figure 3.8. Differential element of radius r coaxial with shaft of radius R with sector $r\Delta\theta$ shown.

Imagine now a second differential element mapped onto the surface of the cylindrical differential element, with sides $r\Delta\theta$ and Δx. The curved sides of this element have (arc) length $r\Delta\theta$, but you can imagine that if the element is small enough ($\Delta\theta$ very small), the "curve" is essentially a straight line. The original cylindrical shaft (and associated differential elements) is subjected to a twisting moment (torque) M such that the cross sections at x and $x + \Delta x$ undergo rotations of ϕ and $\phi + \Delta\phi$, respectively (Figure 3.9a).

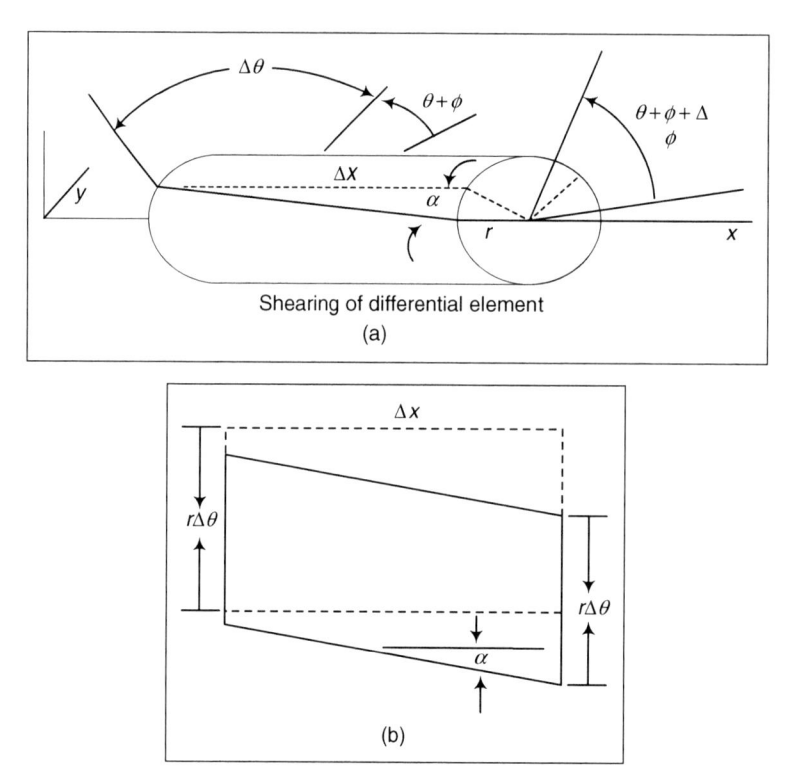

Figure 3.9. Differential elements of the shaft under the action of a torque M: (a) Isometric view and (b) plan view. Note that the originally rectangular differential element $r\Delta\theta$ by Δx has been <u>sheared</u> into a parallelogram.

The *shearing strain* of the surface element is given by the small angular distortion α (recall that $\varepsilon_r = \varepsilon_\theta = 0$, so that the differential elements do not change size). From Figure 3.9 using the small angle approximation:

$$\alpha \approx \tan \alpha = \frac{r\Delta\varphi}{\Delta x} \tag{3.10}$$

Then the shear strain $\gamma_{x\theta}$ (the only nonzero strain) by definition is

$$\gamma_{x\theta} = \lim_{\Delta x \to 0} \alpha = \lim_{\Delta x \to 0} \frac{r\Delta\varphi}{\Delta x} \tag{3.11a}$$

or

$$\gamma_{x\theta} = r\frac{d\phi}{dx} \tag{3.11b}$$

This is the *shear strain–(angular) displacement* relation. The shear strain is a function of radial position r and *twist gradient* $d\phi/dx$. The maximum shear strain thus exists at the surface (where r is maximum, i.e., $r = R$) for any twist gradient and for any radius wherever the twist gradient is maximum. (For a uniform torque on a shaft, the twist gradient is constant. Why?)

3.3 CONSTITUTIVE MODELS

Events in continuum mechanics are governed by three types of equations that must be solved simultaneously; these are (1) the equations of motion (dynamic) or the equilibrium equations (specialized for statics), (2) the kinematic relations, and (3) the constitutive laws. The equilibrium equations and the kinematic relations are general, being directly derived from physics. The kinematic relations connect displacements to the deformation quantities, i.e., they are simply geometric relations. The equilibrium and kinematic relations, together with the conservation of mass and energy, form the balance laws of continuum mechanics.

Since the balance laws make no differentiation with regard to different materials, it is not surprising that these equations are insufficient to describe completely the response of a material. The character of the material is brought into the formulation through the appropriate *constitutive equation* for each material.

A constitutive equation defines an ideal material. There is no definite rule that includes all of the variables that must be present for a given material. Thus, a constitutive theory is intended to describe a limited number of physical phenomena decided at the outset for a given material.

Any constitutive relation must satisfy a number of theoretical requirements. In addition, there are practical considerations as well. For example, any constitutive relationship must be measurable, i.e., one must be able to measure the constitutive parameters that the equations depend upon. Moreover, one typically has to extrapolate from simple test loadings to response under complex stress states. Also, short-time behavior must often be extrapolated to long-time response.

Solid biological materials range from intrinsically stiff to compliant; however, we will learn that compliant materials dominate living systems. Intrinsically stiff biomaterials include bone, cartilage, wood, and nacre—a linear elastic material model might be a good choice here. Examples of compliant materials include cell membranes, muscles, tissue, and various biopolymers, with nonlinear elasticity and viscoelasticity being the useful constitutive models.

3.3.1 Linear Elasticity

Material response is considered elastic if upon removal of the loading the material returns identically to its unloaded state, which implies that no work of loading is dissipated. For an elastic material, the "observable variables" (chiefly strain and temperature) are sufficient to characterize the "state." (For an inelastic material, additional internal state variables are required. Since these are process dependent, they give a *hereditary* nature to the constitutive relation, i.e., the current state depends on the state history.)

In an elastic material, loads and associated deformations may be linearly or nonlinearly related. (A linear elastic material is also called a "Hookean" material after the great mechanician Robert Hooke.) In the theory of linear elastic homogeneous materials, shear or *deviatoric* effects (those relating to angular distortions of differential elements) are uncoupled (independent) from *dilatational* (volume changes in elements) effects associated with normal stresses and strains. In practice, this means that normal stresses are only related to normal strains, and shear stresses are only related to shear strains. If the material is also isotropic, the Hookean constitutive relations are given by

$$\begin{Bmatrix} \sigma_x \\ \sigma_y \\ \sigma_z \end{Bmatrix} = \frac{E}{(1+\nu)(1-2\nu)} \begin{bmatrix} 1-\nu & \nu & \nu \\ \nu & 1-\nu & \nu \\ \nu & \nu & 1-\nu \end{bmatrix} \begin{Bmatrix} \varepsilon_x \\ \varepsilon_y \\ \varepsilon_z \end{Bmatrix} \quad (3.12a)$$

$$\begin{Bmatrix} \tau_{xy} \\ \tau_{yz} \\ \tau_{xz} \end{Bmatrix} = \begin{bmatrix} G & 0 & 0 \\ 0 & G & 0 \\ 0 & 0 & G \end{bmatrix} \begin{Bmatrix} \gamma_{xy} \\ \gamma_{yz} \\ \gamma_{xz} \end{Bmatrix} \quad (3.12b)$$

where E and G are the elastic and shear modulus, respectively, and ν is the Poisson's ratio. (Poisson's ratio provides a measure of the deformation in (say) the y direction when a material is strained in the (say) x direction. For an isotropic material, ν is the same in all directions, typical values ranging from about 0.2 to 0.5.) A loading–unloading diagram for a linear elastic material is shown in Figure 3.10.

The atomistic basis for elasticity is important to understand for at least two reasons. For one, it allows the engineer to view elastic material properties from a Design Degree of Freedom perspective. For another, the concept of material response related to internal energy is the basis

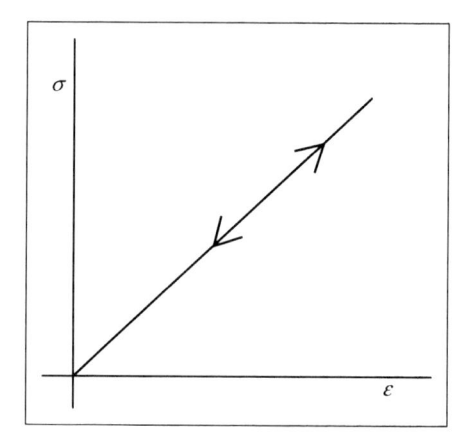

Figure 3.10. Loading–unloading curve for a linear elastic material.

for nonlinear elasticity, which we take up next. Unfortunately, the atomistic study is outside the scope of this chapter and we must refer the reader elsewhere (see, e.g., Jenkins and Khanna, 2005).

3.3.2 Nonlinear Elasticity

A structure may respond nonlinearly in one (or both) of two ways:

1. Nonlinear geometric deformation
2. Nonlinear material response

We will cover nonlinear geometric deformation in Chapter 5, Compliant Structures. Here we focus on nonlinear material response, in particular, nonlinear elasticity. The loading–unloading curve for a nonlinear elastic material response might be as shown in Figure 3.11.

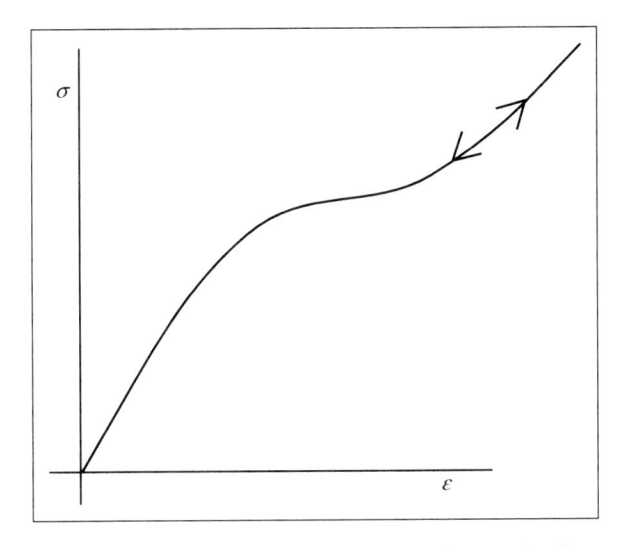

Figure 3.11. Nonlinear elastic material loading–unloading curve.

It is quite common that nonlinear elastic response is also associated with large (finite) deformations, i.e., strains larger than the infinitesimal strains of the linear theory. A hyperelastic constitutive model is often used for nonlinear elastic, finite deformation materials. A material is said to be *hyperelastic* if it possesses a homogeneous stress-free state, and if there exists a *strain energy density function* W, which is an analytic function of the strain tensor (essentially, that it is representable by a convergent power series), with the work done by the stresses equaling the gain in strain energy:

$$s_{ij} = \frac{\partial(W)}{\partial e_{ij}} \qquad (3.13)$$

where e_{ij} and s_{ij} are large deformation strain and stress components, respectively, and W is the strain energy per unit volume of the material. The stress and strain must form a *work conjugate* pair. Common examples of conjugate stress–strain pairs are:

1. Second Piola–Kirchhoff stress ⟷ Green–Lagrange strain
2. Cauchy stress ⟷ Eulerian strain

The concept of "stretch" is also used in large deformation analysis (Figure 3.12).

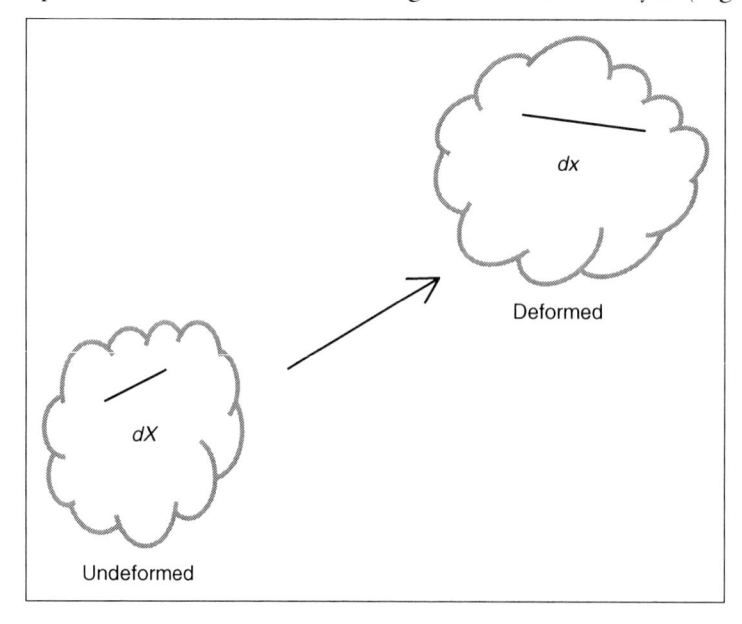

Figure 3.12. The concept and definition of stretch.

For our purposes, for a 1D deformation:

$$\lambda = \frac{deformed\ length}{undeformed\ length} = \frac{dx}{dX} \quad (3.14a)$$

then

$$\varepsilon = \frac{dx - dX}{dX} = \lambda - 1 \quad (3.14b)$$

Incompressible materials have $\nu = 0.5$ and undergo no volume change (*isochoric*) during deformation. Elastomers are common examples. Isochoric deformation can be described by $\lambda_1 \lambda_2 \lambda_3 = 1$ (where $\lambda_1, \lambda_2, \lambda_3$ are the principal stretches).

Since $\lambda = \varepsilon + 1$, $\partial \lambda = \partial \varepsilon$ and

$$s_{ij} = \frac{\partial(W)}{\partial \lambda_{ij}} \quad (3.15)$$

Examples of common strain energy density functions include the neo-Hookean, Mooney–Rivlin, Blatz–Ko, and Ogden forms.

Example 3.2

For uniaxial deformation of a rubber bar,

$$\lambda_x = \lambda_1 = \lambda, \quad \lambda_2 = \lambda_3 = \frac{1}{\sqrt{\lambda}}$$

and the Mooney–Rivlin strain energy density function is

$$W = C_1\left(\lambda^2 + \frac{2}{\lambda} - 3\right) + C_2\left(\frac{1}{\lambda^2} + 2\lambda - 3\right)$$

Then

$$\sigma_x = \frac{\partial W}{\partial \lambda} = C_1\left(2\lambda - \frac{2}{\lambda^2}\right) + C_2\left(\frac{-2}{\lambda^3} + 2\right) = \cdots = 2(C_1\lambda + C_2)\left(1 - \frac{1}{\lambda^3}\right)$$

Note that if we rewrite the above as

$$\frac{\sigma_x}{2\left(1 - \frac{1}{\lambda^3}\right)} = \lambda C_1 + C_2,$$

we recognize a linear equation in C_1 and C_2 for a given λ, where C_1 is the slope and C_2 is the "y intercept." This can be used to fit the strain energy density test data to the Mooney–Rivlin model (see Problems at the end of this chapter).

3.3.3 Linear Viscoelasticity

The range of material behavior is bounded by the ideal solid and the ideal fluid (see Figure 3.13). In between lie more complex materials, among them are the viscous fluid and the viscoelastic solid.

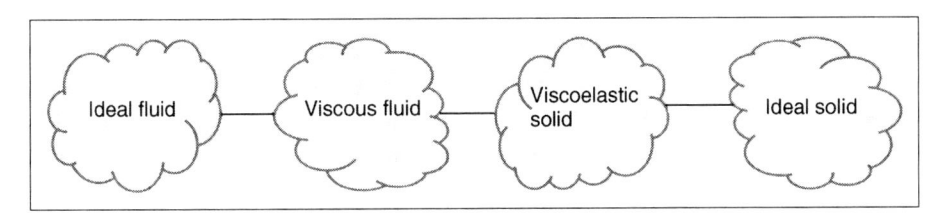

Figure 3.13. Range of material behavior.

The viscous fluids and viscoelastic solids bring the dimension of time into the material's response. In fact, the distinction between solid and fluid is linked to the time scales of observation and loading. If materials have some inherent "time constant," then the apparent response varies with the loading and observation time scales. When the inherent time constant and the observation time are not widely different, viscoelastic material response can be observed. Characteristics of viscoelastic materials include the following:

1. Creep (strain): Creep is a time-dependent deformation in response to a constant load. The three stages of creep are shown in Figure 3.14. (It is common in

Chapter 3

viscoelastic theory to use stress σ as the load. Clearly, if the cross-sectional area changes under constant force, the stress will not be constant. However, linear viscoelasticity is a small deformation theory and as such the force and resulting stress can be considered constant.)

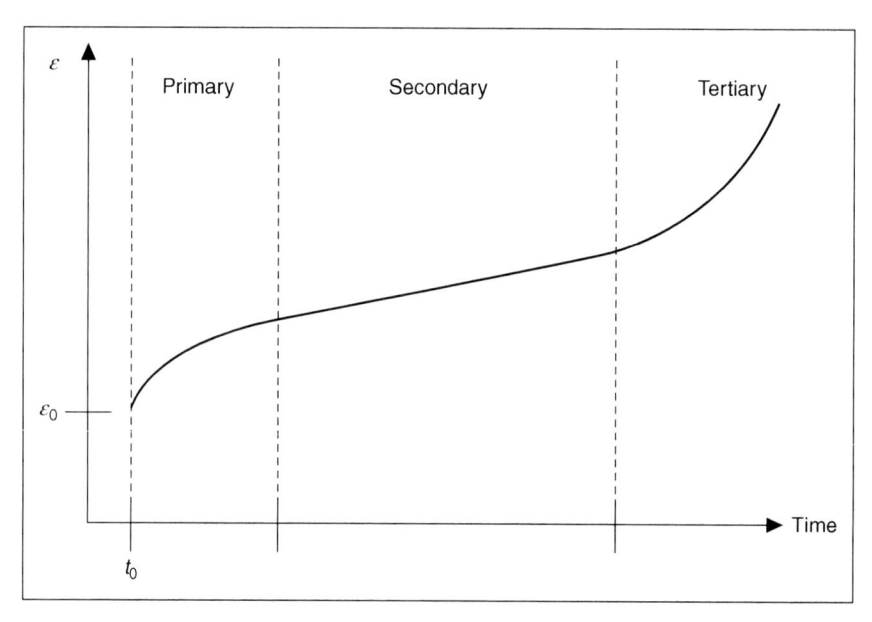

Figure 3.14. The three stages of creep. http://commons.wikimedia.org/wiki/File:3StageCreep.svg

2. Recovery: Following load removal, the instantaneous elastic strain is recovered completely and instantaneously, while the creep strain is recovered to some level of completeness in "delayed" time (see Figure 3.15).

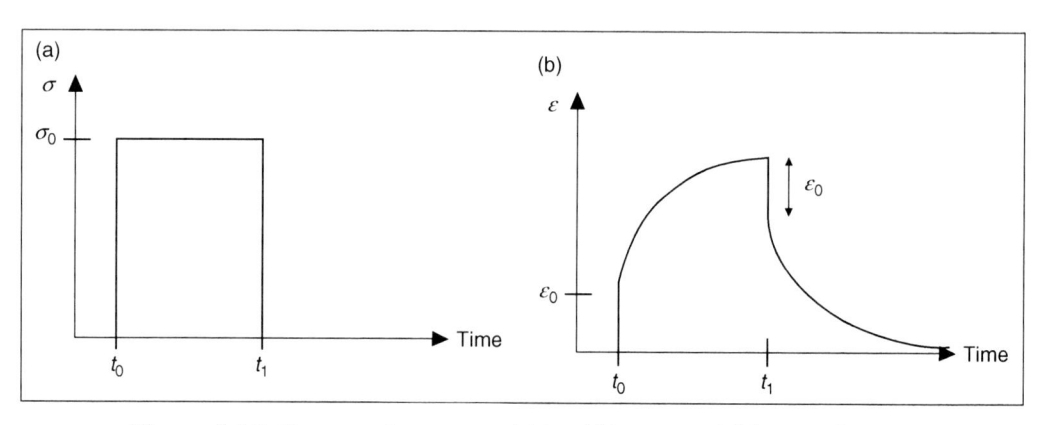

Figure 3.15. Creep and recovery: (a) load history and (b) strain history.
http://commons.wikimedia.org/wiki/File:Creep.svg

3. Relaxation (stress relaxation): Relaxation is a time-varying stress in response to a constant deformation (see Figure 3.16).

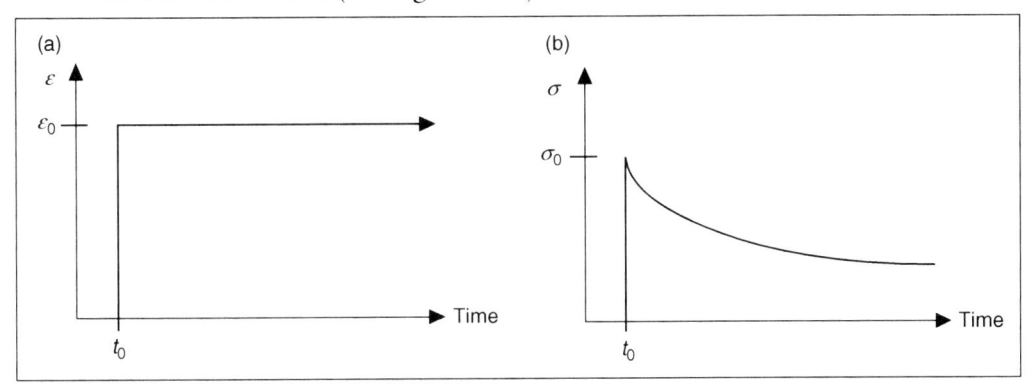

Figure 3.16. Stress relaxation: (a) load history and (b) stress history.
http://commons.wikimedia.org/wiki/File:StressRelaxation.svg

4. Time–temperature superposition: It is readily observed that increasing the specimen temperature above a reference temperature during a creep test accelerates the response, i.e., the final state (say *creep rupture*) is reached sooner than without the temperature increase. That is, the effect of temperature on the time-dependent mechanical behavior is equivalent to a stretching (compressing) of the real time for temperatures above (below) the reference temperature. In other words, the effects of time and temperature may be superposed.

5. Mechanisms of creep: The mechanisms responsible for creep in metals and nonmetals (e.g., polymers) are different. In metals, dislocations → irreversible creep, while in polymers, chain unfolding → reversible creep. Given that there are few metallic materials in biology, but a vast array of biopolymers, it is this latter group that we focus on here.

3.3.4 Linear Viscoelastic Constitutive Models

Models for linear viscoelastic material/structural systems fall into two categories: mechanical analog models and hereditary models. In this chapter, we focus on the former, the analog models (see, e.g., Roylance (1996) for the integral model).

The Kelvin–Voigt Solid

A viscous dashpot (damper) can be modeled as shown in Figure 3.17.

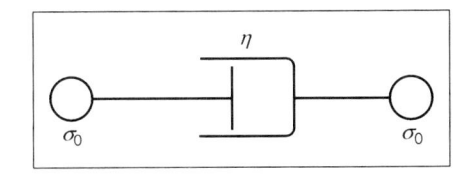

Figure 3.17. A viscous dashpot.

Chapter 3

The dashpot's response to stress σ_0 is Newton's law of viscosity $\dot{\sigma} = \eta\dot{\varepsilon}$, where η is the *coefficient of viscosity* (SI unit is Pa s). Values of η are $\sim 10^{-3}$ Pa s for water, 6 Pa s for honey, and $>10^{12}$ Pa s for solid polymers. Integrating once gives

$$\int d\varepsilon = \frac{\sigma_0}{\eta} \int dt$$

$$\varepsilon(t) = \frac{\sigma_0}{\eta}t + C \qquad (3.16)$$

where $C = 0$, since $\varepsilon = 0$ at $t = 0$.

We can put the dashpot in parallel with an elastic spring (Figure 3.18).

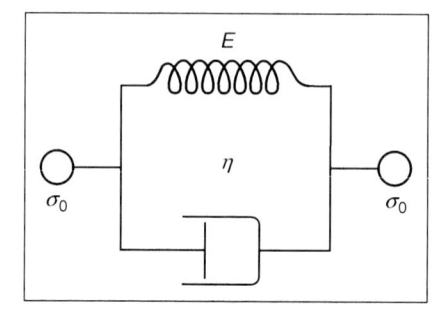

Figure 3.18. A viscoelastic solid model. http://commons.wikimedia.org/wiki/File:Kelvin_Voigt_diagram.svg

Under constant load σ_0, an FBD and an equilibrium statement would give

$$\sigma_0 = \sigma_E + \sigma_\eta$$
$$= E\varepsilon + \eta\dot{\varepsilon}$$

(Note that formally we enforce *force* equilibrium but for small uniaxial deformations we can substitute stresses for forces.) Upon rewriting

$$\frac{d\varepsilon}{dt} + \frac{E}{\eta}\varepsilon = \frac{\sigma_0}{\eta} \qquad (3.17)$$

The solution to equation (3.17) can be shown to be (see Problems at the end of this chapter)

$$\varepsilon(t) = \frac{1}{E}\left(1 - e^{(E/\eta)t}\right)\sigma_0 = J(t)\sigma_0 \qquad (3.18)$$

where $J(t)$ is again the *Creep Compliance*:

$$J(t) = \frac{1}{E}\left(1 - e^{(E/\eta)t}\right) \qquad (3.19)$$

The model is known as the *Kelvin–Voigt Solid*. Note that in this model $\varepsilon(0) = 0$ and thus there is no instantaneous elasticity (see Figure 3.19).

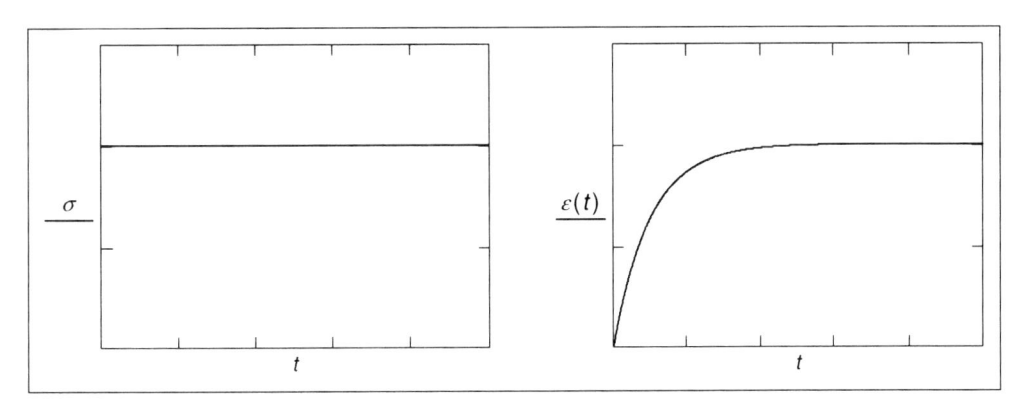

Figure 3.19. Creep response for the Kelvin–Voigt Solid.

Maxwell Fluid

We could also combine the spring and damper in series (see Figure 3.20).

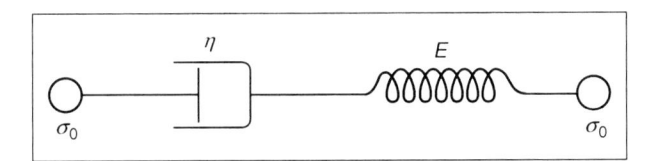

Figure 3.20. Spring and damper connected in series.
http://commons.wikimedia.org/wiki/File:Maxwell_diagram.svg

An FBD and equilibrium statement would give $\sigma_E = \sigma_\eta$ or

$$\varepsilon = \varepsilon_E + \varepsilon_\eta$$
$$\dot{\varepsilon} = \dot{\varepsilon}_E + \dot{\varepsilon}_\eta \qquad (3.20)$$
$$\dot{\varepsilon} = \frac{\sigma}{E} + \frac{\sigma}{\eta}$$

For $\varepsilon = \varepsilon_0$ at $t = 0 \to \dot{\varepsilon} = 0$ (strain instantaneously applied). Then the parallel spring-damper is called the *Maxwell fluid*. In this case, equation (3.20) can be directly integrated to give

$$\sigma(t) = \sigma_0 e^{(E/\eta)t} = E e^{(E/\eta)t}\varepsilon_0 = E(t)\varepsilon_0 \qquad (3.21)$$

where $E(t) = E e^{(E/\eta)t}$ is the *relaxation modulus*. Figure 3.21 shows the relaxation response.

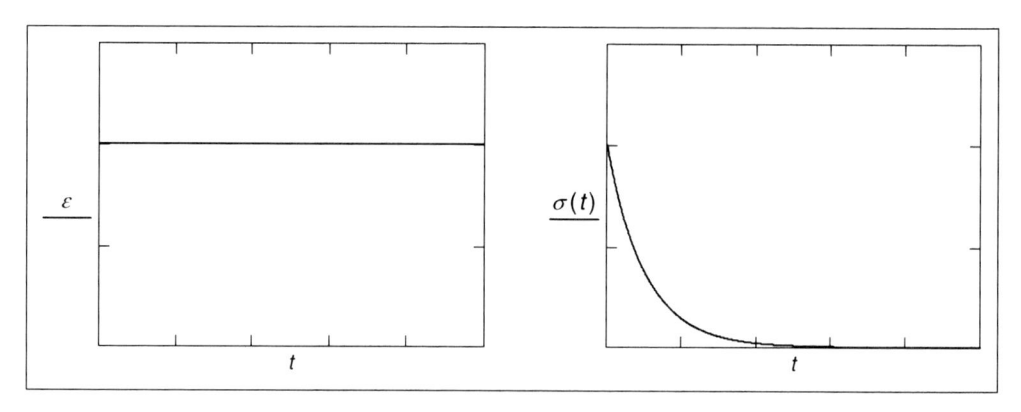

Figure 3.21. Stress relaxation in the Maxwell model.

3.3.5 The Standard Linear Solid

Perhaps the simplest model that reproduces realistic viscoelastic material behavior is the *Standard Linear Solid* (see Figure 3.22).

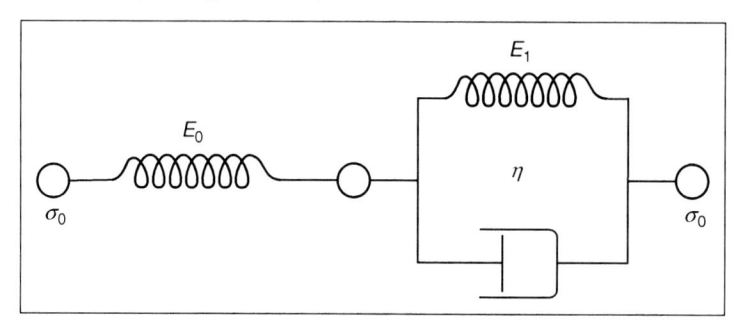

Figure 3.22. The concept of the Standard Linear Solid.
http://commons.wikimedia.org/wiki/File:Kelvin_Voigt_diagram.svg

Note that for spring E_0, $\varepsilon_0 = \sigma/E_0$. For the Kelvin–Voigt group, recall

$$\sigma = E\varepsilon + \eta\dot{\varepsilon} \qquad (3.22)$$

The total strain is $\varepsilon = \varepsilon_0 + \varepsilon_1$ and the total strain rate is $\dot{\varepsilon} = \dot{\varepsilon}_0 + \dot{\varepsilon}_1$. Substituting into equation (3.22) gives (this is left as an exercise for the student):

$$\left(\frac{1}{E_1} + \frac{1}{E_2}\right)\sigma + \frac{\eta}{E_1 E_2}\dot{\sigma} = \varepsilon + \frac{\eta}{E_1}\dot{\varepsilon} \qquad (3.23)$$

Let $a = (E_0 + E_1)/\eta$, $b = E_1/\eta$, then equation (3.23) can be written as

$$\dot{\sigma} + a\sigma = E_0(\dot{\varepsilon} + b\varepsilon) \qquad (3.24)$$

Note that if $\sigma = \sigma_0$ = constant (which implies that $\dot\sigma = 0$) and the load acts for a very long time (so that $\dot\varepsilon \to 0$):

$$\sigma_0 = \frac{b}{a}E_0\varepsilon = \frac{E_0 E_1}{E_0 + E_1}\varepsilon \qquad (3.25)$$

The quantity bE_0/a is the *long-term modulus*. Now if $\sigma = \sigma_0$, equation (3.24) becomes

$$\dot\varepsilon + b\varepsilon = \frac{a}{E_0}\sigma_0 \qquad (3.26)$$

The solution to equation (3.26) can be shown to be (see Problems at the end of this chapter)

$$\varepsilon(t) = \frac{\sigma_0}{E_0} + \frac{\sigma_0}{E_1}(1 - e^{(E_1/\eta)t}) \qquad (3.27)$$

A plot of the solution (3.27) is shown in Figure 3.23.

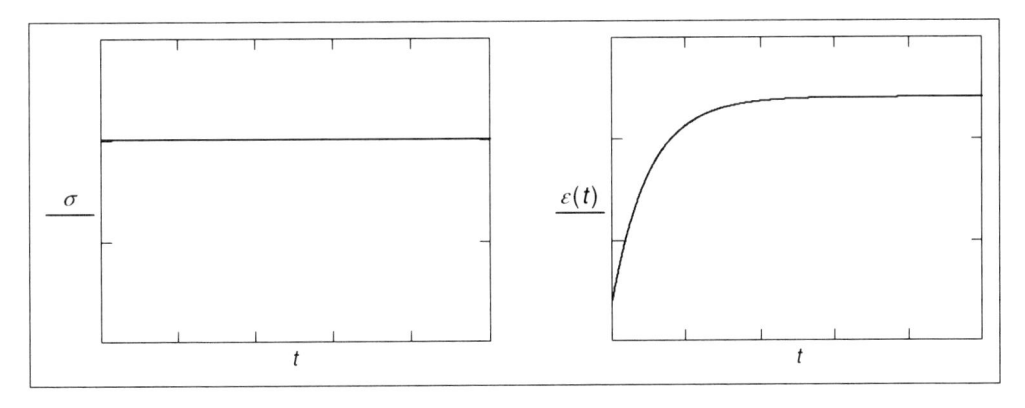

Figure 3.23. Plot of the Standard Linear Solid creep response.

If the load is removed at time $t = t_1$, the creep strain is given by

$$\varepsilon(t) = \frac{\sigma_0}{E_0}\frac{a-b}{b}(e^{-b(t-t_1)} - e^{-bt}), \quad t > t_1 \qquad (3.28)$$

Note that $\sigma_0 = \varepsilon_0 E_0$ is the instantaneous elasticity and E_0 is the "*short-term*" or Young's modulus.

Going back to equation (3.26) but with $\dot\varepsilon = 0$, we can model relaxation behavior with the Standard Linear Solid:

$$\dot\sigma + a\sigma = bE_0\varepsilon \qquad (3.29)$$

With $\sigma_0 = E_0\varepsilon_0$, the solution to equation (3.29) is

$$\sigma(t) = \frac{E_0 E_1}{E_0 + E_1}\left[1 - \frac{E_0}{E_1}e^{(E_0 + E_1)t/\eta}\right]\varepsilon_0 \qquad (3.30)$$

where the relaxation time constant is given by $\eta/(E_0 + E_1)$. Combinations of Kelvin–Voigt and Maxwell models can represent more complex material behavior.

Example 3.3

Polyethylene has a short-term elastic modulus of 149 MPa (21.6 ksi) and a long-term modulus of 78.3 MPa (11.9 ksi). η may be taken as 1.2×10^{12} Pa s.

(a) Find E_1

$E_0 = 149$ MPa, long-term modulus $= (E_0 E_1)/(E_0 + E_1) = 78.3$ MPa $\Rightarrow E_1 = 165$ MPa

(b) Creep time constant

$\eta/E_1 = 1.2 \times 10^{12}$ Pa s$/165 \times 10^6$ Pa $= 7{,}266$ s $= 2.02$ h

(c) Relaxation time constant

$\eta/(E_0 + E_1) = 3{,}821$ s $= 1.06$ h

KEY POINTS TO REMEMBER

- *Structures* are physical artifacts that carry loads other than their own weight.
- Every *structure* is a *material* as well. Hence the *mechanics* of a structure is intimately linked with the properties of the structural material.
- Critical elements of structural mechanics include geometry, loads, boundary conditions, and load paths.
- Critical elements of structural analysis include configuration, equilibrium, deformation, and constitution.
- Biological materials exhibit complex behavior such as nonlinear elasticity or linear viscoelasticity.

REFERENCES

Boley, B. A., and J. H. Weiner. 1997. *Theory of Thermal Stresses*. Dover.
Jenkins, C. H., and S. Khanna. 2005. *Mechanics of Materials: A Modern Integration of Mechanics and Materials in Structural Design*. Elsevier.
Roylance, D. 1996. *Mechanics of Materials*. Wiley.
Schodek, D. L. 2001. *Structures*, 4th ed. Prentice-Hall.

PROBLEMS

3.1. A round bar of steel, 9 in. long with a diameter of 0.5 in., is subjected to a tensile force of 10,000 lb. What are the dimensions of the bar after the stress is applied? Provide an answer in both the US system of units and the SI system. Elastic modulus of steel is 30×10^6 psi and Poisson's ratio is 0.3.

3.2. A bridge is 1 km long and is made of steel. The temperature of the bridge varies from -20 to $40°C$. How much thermal expansion results? If the joints in the bridge cannot accommodate the expansion, what is the stress in the bridge at the lowest and maximum temperature if the bridge was erected at a temperature of $10°C$? The thermal expansion coefficient of steel is 6×10^{-6} mm/mm/°C over this temperature range.

3.3. Three rods are hung vertically from a ceiling, each suspended from their top. Each rod is 10 m in length, with one made from steel, another from aluminum, and the third from wood. Compute the total elongation of each rod due to its own weight, and compare one to another. Why are the results independent of the cross-sectional area?

Material	Density (kg/m³)	Modulus (GPa)
Steel	7,850	200
Aluminum	2,700	70
Wood	500	12

3.4. A helicopter needs to lift a 20-ton tank that is resting on the ground. A 300-ft titanium cable of 1 in. diameter is lowered and connected to the tank. Assuming the tank lifts uniformly, how much elevation must the copter gain before the tank is lifted free from the ground?

3.5. A bar of length L, elastic modulus E, and density ρ is to be hung vertically from the ceiling. A tensile force P is to be applied axially at the lower end. Derive an expression for the total elongation of the bar under the action of the applied force plus the self-weight of the bar. Under what conditions can the self-weight of the bar be neglected?

3.6. A table leg has an "inverted taper" as shown:

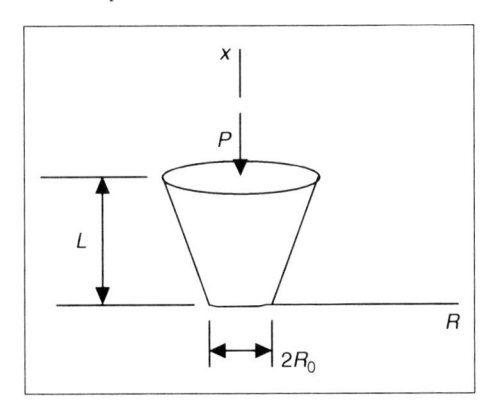

a. Show that the "taper" can be modeled by

$$R = R_0\left(c\frac{x}{L} + 1\right)$$

where c is a constant.

b. Assuming small taper ($c \ll 1$), determine an expression for $\sigma_x(x)$.

c. Assuming small taper ($c \ll 1$), determine an expression for the total axial deflection u_{total}.

3.7. A 32 in. long by 1.6 in. diameter solid prismatic circular shaft is made of an aluminum alloy that has an allowable shear stress (strength) of 10 ksi, and a shear modulus of elasticity $G = 3{,}800$ ksi. If the allowable angle of twist over the 32 in. length of the shaft is 0.10 rad, what is the value of the allowable end torque? (Hint: you must investigate from both strength and stiffness perspectives, then compare.)

3.8. A 2 m long solid prismatic circular shaft is made of brass that has an allowable shear stress (strength) of 120 MPa and a shear modulus of elasticity $G = 39$ GPa. Over the shaft length, the angle of twist allowed is 0.10 rad. If the shaft is subjected to a maximum end torque of 25 kN m, what is the required minimum diameter of the shaft? (Hint: you must investigate from both strength and stiffness perspectives, then compare.)

3.9. A uniaxial tension test performed on an elastomer generated the force-deflection data provided. The elastomer specimen had an initial gauge length of 0.491 in. and an undeformed cross-sectional area of 1 in.^{2}. Using linear regression, determine a set of Mooney–Rivlin coefficients to model the material response.

Displacement (in.)	Force (lb)
8.95E–04	7.33
5.45E–03	41.5
8.06E–03	75.7
1.13E–02	129
1.44E–02	183
1.80E–02	252
2.55E–02	393
2.94E–02	462
3.28E–02	535
4.00E–02	672
4.39E–02	740
4.75E–02	813
5.46E–02	945
5.80E–02	1,010
6.85E–02	1,150
7.30E–02	1,280
7.69E–02	1,350
8.05E–02	1,420
8.80E–02	1,560
9.19E–02	1,630
9.55E–02	1,700
1.03E–01	1,860
1.06E–01	1,930
1.17E–01	2,200
1.28E–01	2,500
1.38E–01	2,850
1.50E–01	3,260
1.60E–01	3,730
1.71E–01	4,220
1.77E–01	4,420
1.83E–01	4,480

Chapter 3

3.10. For the standard linear viscoelastic solid model:

a. Derive the governing differential equation:

$$\dot{\varepsilon} + b\varepsilon = \frac{a}{E_0}\sigma_0$$

b. Show that the solution is

$$\varepsilon(t) = \frac{\sigma_0}{E_0} + \frac{\sigma_0}{E_1}\left(1 - e^{(E_1/\eta)t}\right)$$

3.11. Show that

a. The creep response for the Kelvin–Voigt model is given by

$$\varepsilon = \frac{1}{E}\left[1 - e^{-(E/\eta)t}\right]\sigma_0 = J(t)\sigma_0$$

b. The stress relaxation response for the Maxwell model is given by

$$\sigma = \left[Ee^{-(E/\eta)t}\right]\varepsilon_0 = E(t)\varepsilon_0$$

CHAPTER 4

Structures and Materials in Nature

The harmony of the world is made manifest in Form and Number.

—*D'Arcy Thompson*

In the previous chapter, we reviewed some specific topics in engineering structures and materials germane to natural structures and materials. In the present chapter, we will study selected topics in natural structures and materials, starting with natural configurations (size and shape), then exploring shape and structure, and concluding with some important fundamentals of natural materials.

4.1 CONFIGURATION IN NATURE

4.1.1 Size—Background

The range of size in biology is vast. The single cells such as protozoa may be a few tenths of micrometers in characteristic length, while the blue whale can grow to 25 m in length. This range covers about eight orders of magnitude and does not take into account the size of critical biological components such as DNA or lakes and oceans.

The fossil record reveals two facts of interest here. First, the oldest rocks contain only fossils of minute bacteria. Second, the range of size present at any snapshot in time has increased over time. One would conclude from these observations that natural selection favors increase in size and thus there are advantages to being large. This is surprising, on the one hand, since increase in size also represents increase in complexity and increased time for development. On the other hand, increased complexity improves options for survival, for finding alternatives to environmental pressure. (There is evidence that some species today have evolved smaller in size from larger ancestors; the hummingbird is an example.)

Environmental pressures include short-term factors such as wind, and longer-term, but still transient, factors such as drought and food supply. The one constant factor, always present, is gravity. Gravity may be the single most dominant factor affecting biological size (and shape), and we will discuss more about gravity later.

The concept of *scaling* is used in biology in a similar fashion to engineering. For example, we know that geometric perimeter "scales" with length ($C \sim L$), area "scales" with the square of an object's length ($A \sim L^2$), and volume "scales" with length cubed ($V \sim L^3$) (see Figure 4.1). This is a form of isometric scaling that we will explore next as part of scaling in nature.

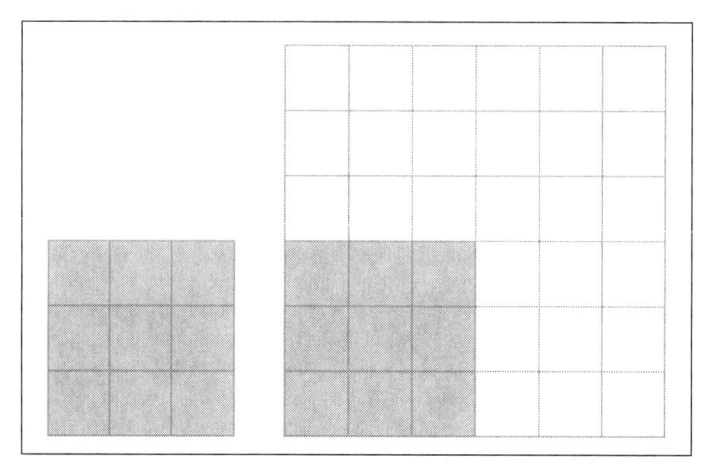

Figure 4.1. Geometric scaling. Doubling the length increases the area by four times.

4.1.2 Allometry

Isometry or *isometric scaling* occurs when proportions do not change with changes in size (either from growth/development or evolutionary changes). The geometric scaling discussed earlier is an example. Isometry is a special case of *allometry*. Think of isometry as "zooming" on a copier.

Any deviation from isometry is called *allometry* or *allometric scaling*. The standard allometric relationship for representing changes in proportion is

$$y = bx^a \qquad (4.1)$$

Equation (4.1) represents the special case of isometry where

$a = 1$ when $x =$ length, $y =$ length
$a = 2$ when $x =$ length, $y =$ area
$a = 1$ when $x =$ length, $y =$ volume
$a = \frac{1}{3}$ when $x =$ mass, $y =$ length
$a = \frac{2}{3}$ when $x =$ mass, $y =$ area
$a = 1$ when $x =$ mass, $y =$ volume

For all other values, equation (4.1) represents allometry. Equations of the form (4.1) will plot as straight lines on a log–log plot (see Figure 4.2):

$$y = bx^a$$
$$\log y = \log(bx^a) = \log b + \log(x^a) = a \log x + \log b$$

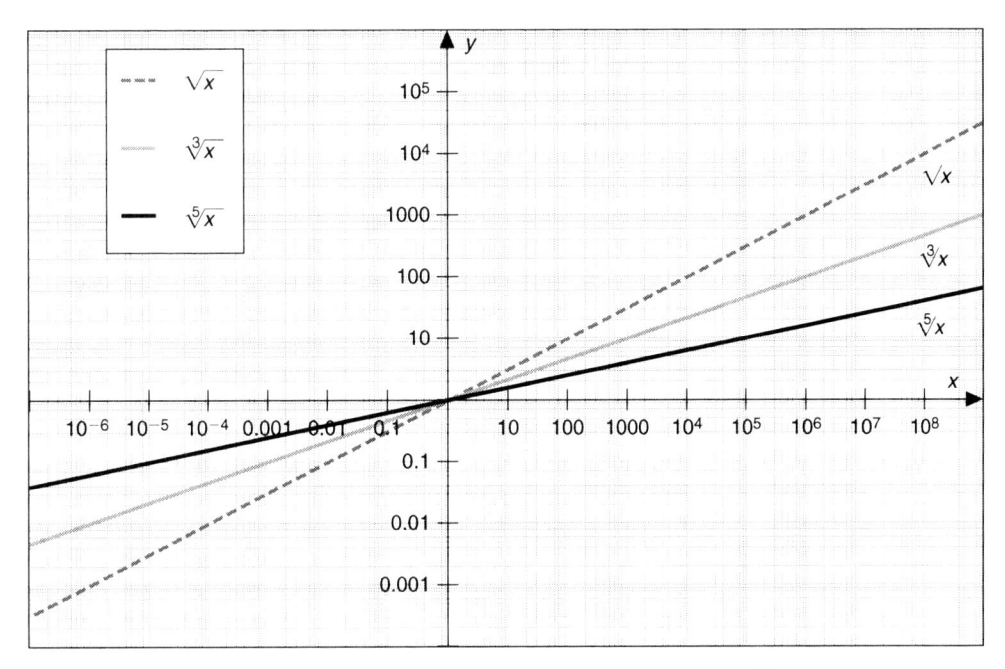

Figure 4.2. Log–log plot of equation (4.1). The function $y = \sqrt{x}$ has a slope of ½ (e.g., $\log(10/1)/\log(100/1) = ½$). The function $y = x^{1/5}$ has a slope of ⅕, etc. (Verification of these slopes is left as an exercise for the student.)

A simple example of isometry in biology is the relationship of arm span to height in humans. Data plotted on a log–log plot shows a slope $a = 1$ for older ages. (Interestingly, for babies younger than approximately 1 year, the relationship of arm length to body height deviates slightly from isometry.)

Natural selection may lead to structure–function relationships that optimally meet the demands placed on the organism. If organisms do not change their proportions as they change their size, then function may be altered, which drives evolutionary processes.

Example 4.1

Consider that globular, spherical-form organisms larger than approximately 1 mm in size are not readily found in nature. The surface area (A_s) and volume (V) of a sphere are, respectively

$$A_s = 4\pi R^2, \quad V = \frac{4}{3}\pi R^3$$

Chapter 4

or

$$\frac{A_s}{V} = \frac{4\pi R^2}{(4/3)\pi R^3} = \frac{3}{R} = \frac{6}{D}$$

At 1 mm diameter, $A_s/V = 6/1$ mm $= 6$ mm^{-1}.

At 1 m diameter, $A_s/V = 6/1{,}000$ mm $= 6 \times 10^{-3}$ mm^{-1}, i.e., 1,000 times smaller!

Hence, if gas transfer, heat transfer, and other similar processes that depend on the surface area to volume ratio are adequate for the spherical form at 1 mm diameter, they would be inadequate at 1 m diameter. In order to scale up in size, a different configuration would be needed, one that changes proportion with size, namely an allometric scaling.

Example 4.2

An example of allometric scaling is the egg mass to adult mass in birds. Table E4.1 provides some representative data.

Table E4.1. Selected values of egg mass to adult mass in birds (after Calder, 1984)

Species	Adult Mass (g)	Egg Mass (g)
Hummingbird	3.6	0.6
Pigeon	283	14
Pheasant	1,020	34
Duck	3,629	94.5
Goose	4,536	165.4
Ostrich	113,380	1,700

From a log–log plot (see Figure E4.1), the mass relationship is

$$\log(m_{egg}) = \log(0.198) + 0.77 \log(m_{adult})$$

or

$$m_{egg} = 0.198\, m_{adult}^{0.77}$$

The exponent is approximately ¾ and indicates that the egg mass doesn't increase as fast as adult mass. (An exponent of 1 would mean that the egg mass and the adult mass were proportional, while an exponent greater than 1 would mean that the egg mass increased faster than the adult mass.)

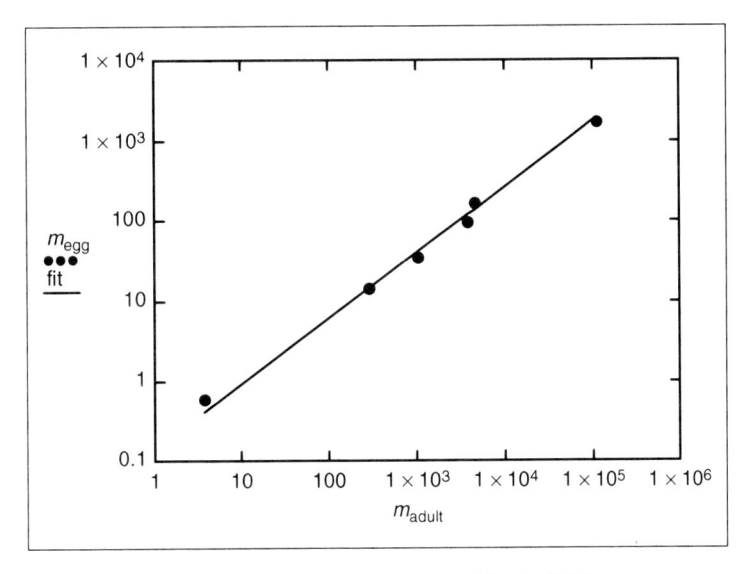

Figure E4.1. Log–log plot of Table E4.1.

Example 4.3

Another example of an allometric scaling is the length (L) to trunk diameter (D) in trees. Data shows that $D \sim L^{3/2}$, that is, the trunk diameter increases faster than does the length. Biomechanical models suggest that it is necessary to maintain a factor of safety against buckling. Greenhill's buckling model for the critical buckling height L_{cr} under self-weight is

$$L_{cr} = \frac{\sqrt{\pi E r_0^2 c(b - 4n + 2)}}{4(m_{total} g)^{0.5}} \tag{4.2}$$

where

r_0 = the basal trunk radius (m)
m_{total} = total biomass (kg)
n = taper parameter
b = biomass distribution
c = Bessel function root

For a tree assumed to be a tapered pole of constant weight density ρ_w and negligible branch and leaf biomass, $b = 2n + 1$. If the trunk is modeled as a cylinder, $n = 0$, $b = 1$, and $c = 1.867$, then

$$L_{cr} = 0.792 \left(\frac{E}{\rho_w}\right)^{1/3} D^{2/3} \tag{4.3}$$

If the elastic modulus and mass density of a spruce tree are taken as $E = 9$ GPa and $\rho = 0.6$ Mg/m^3, the critical length for a 1-m-diameter trunk is

$$L_{cr} = 0.792 \left(\frac{9 \times 10^9 \, \text{N/m}^2}{0.6 \times 10^3 \, \text{kg/m}^3 \times 9.81 \, \text{m/s}^2}\right)^{1/3} (1 \, \text{m})^{2/3} = 91 \, \text{m}$$

4.1.3 Gravity Effects on Scaling in Biology

Because mass $\sim L^3$ but structural strength scales by cross-sectional area $\sim L^2$, the self-weight loading grows faster than the structural resistance to load. Hence, as organisms grow large, a greater proportion of overall mass must be dedicated to structural support. For example:

- An elephant weighing approximately 4 tons cannot run at great speeds, due, in part, to loads placed on joints.
- Redwoods and Sequoias weighing on the order of 6,000 tons have most of the mass dedicated to structure, rather than in muscles and organs.

What happens to life as we know it when gravity changes? Examples of humans and other biological organisms in extended spaceflight are readily available. We know that humans lose bone and muscle mass, water migration in plants is disrupted, and the lungs of tadpoles fail to inflate (Morey-Holton, 2003). There is still much to learn about the effects of gravity on biology.

4.1.4 Fractal Scaling

Fractals were first conceptualized in the 1960s in a study of the length of coastlines. That length was determined to be a function of how closely you looked, due to an infinity of nooks and crannies.

Fractals are used to describe geometries that exhibit *self-similarity*. In a self-similar geometry, at each successive level of magnification, the geometry appears to be the same, to have the same morphology, as at the previous level. A simple example is the Koch "snowflake" (Figure 4.3).

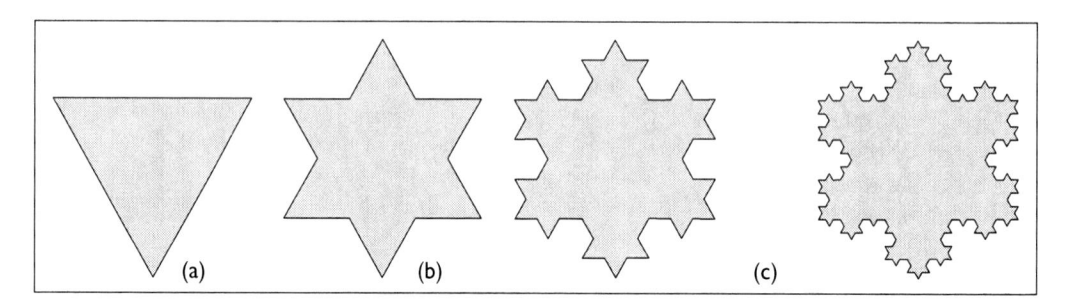

Figure 4.3. Geometry of the Koch snowflake. This geometry has fractal dimension of 1.26. (a) Begin with an equilateral triangle. (b) Replace the middle third of each line segment with a pair of line segments forming an equilateral "bump." (c) Repeat.
http://en.wikipedia.org/wiki/File:Von_Koch_curve.gif.

Other examples of fractals in nature include branching of veins (capillaries, leaves) and trees (Figure 4.4). Lung alveoli (the air sacs wherein exchange of oxygen and carbon dioxide takes place) have been discussed not as an "area" but as a fractal to reconcile competing, but disparate, measurements (Mandelbrot, 1983).

Figure 4.4. Fractal nature of tree branching.
http://commons.wikimedia.org/wiki/File:Maple_Tree_Fractal_Branch_1.jpg.

4.2 SHAPE AND STRUCTURE IN NATURE

4.2.1 Shape

Shape in nature follows the adage "form follows function." Shape is influenced by size, structural support, and energy management, among others. Recall our discussion on the isometric scaling of spheres or spherical forms. Surface area to volume scales by ⅔, and this is problematic for an increase in the size of the organism. The cylinder on the other hand is different; for a long or open cylinder, we have

$$A_s = 2\pi RL$$
$$V = \pi R^2 L \tag{4.4}$$

For a fixed radius, we can increase the size of a cylinder by increasing its length and thus preserve the A_s/V ratio!

With cylinders we can also *branch* (Figure 4.5).

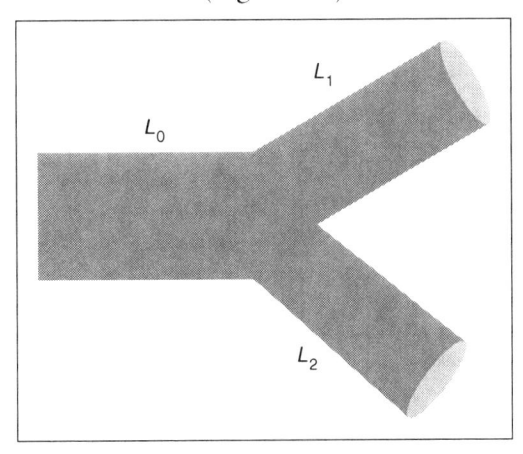

Figure 4.5. Bifurcation branching with cylinders.

Example 4.4

Referring to Figure 4.5, for flow continuity,

$$A_0 = A_1 + A_2$$

where A_0, A_1, and A_2 are the cross-sectional areas of branches L_0, L_1, and L_2, respectively. If $A_1 = A_2$ and $L_1 = L_2$ (*symmetric branching*), then

$$A_0 = 2A_1 \rightarrow A_1 = \frac{A_0}{2}$$

giving

$$\pi R_1^2 = \frac{\pi R_0^2}{2} \Rightarrow R_1 = \frac{R_0}{\sqrt{2}} = R_2$$

Then with $L_1 = L_2 = L_0 = L$,

$$A_{s1} = A_{s2} = 2\pi R_1 L = \frac{2\pi R_0 L}{\sqrt{2}}, \quad A_{s0} = 2\pi R_0 L$$

$$V_1 = V_2 = \pi R_1^2 L = \frac{\pi R_0^2 L}{2}, \quad V_0 = \pi R_0^2 L$$

$$\frac{A_{s,\text{total}}}{V_{\text{total}}} = \frac{A_{s0} + 2A_{s1}}{V_0 + 2V_1}$$

$$= \frac{2\pi R_0 L + 4\pi(R_0/\sqrt{2})L}{\pi R_0^2 L + \pi R_0^2 L}$$

$$= \frac{(2 + (4/\sqrt{2}))R_0}{2R_0^2}$$

$$= \frac{1 + (2/\sqrt{2})}{R_0}$$

Comparing this answer to the initial $A_{s0}/V_0 = 2/R_0$, the change is

$$\frac{(1 + (2/\sqrt{2})/R_0) - (2/R_0)}{2/R_0} = 0.207$$

or a 21% increase in surface area to volume ratio by the symmetric branching.

To further illustrate how function drives form, consider the primary functions of a plant (following Niklas, 1992):

a. Reproduction (spore release, pollen capture, seed and fruit release)

b. Photosynthesis (light interception, gas exchange)

c. Hydraulics (absorption, transport, evaporation)

d. Structural support (static loads, dynamic loads, nastic movement)

Photosynthesis, for example, requires both light interception and gas exchange:

$$6CO_2 + 6H_2O \rightarrow C_6H_{12}O_6 + 6O_2$$
$$\text{Carbon dioxide} \quad \text{Water} \quad \text{Energy} \quad \text{Sugar} \quad \text{Oxygen}$$

Both light interception and gas exchange depend on surface area to volume ratio, that is—shape! Fick's first law describes gas diffusion through a cell wall and can be written in one dimension as

$$\frac{dN}{dt} = -\chi A_s \left(\frac{d\psi}{dx} \right) \tag{4.5}$$

where χ is the diffusion coefficient (m²/s), A_s is the surface area, ψ is the concentration (mol/m³), and N is the amount of substance (molecular species) transported per unit time t. The surface area is explicit in equation (4.5), while the volume comes in through the concentration.

Example 4.5

Determine the minimum surface area of a cylinder with fixed volume.

The volume V_0 and surface area A_s of a closed right circular cylinder of radius R and length L are

$$A_s = 2\pi R L + 2\pi R^2$$
$$V_0 = \pi R^2 L$$

Noting that $V_0/R = \pi R L$ and substituting into the above gives

$$A_s = \frac{2V_0}{R} + 2\pi R^2$$

A plot of A_s vs. R is shown in Figure E.4.2.

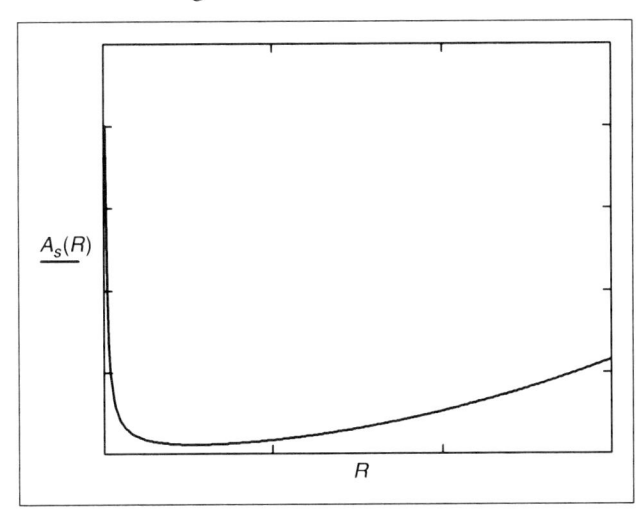

Figure E.4.2. A plot of A_s vs. R for a closed right circular cylinder.

Chapter 4

Now finding the minimum surface area

$$\frac{\partial A_s}{\partial R} = 4\pi R - 2\frac{V_0}{R^2} = 0 \Rightarrow R^3 = \frac{1}{2\pi} V_0$$

Substituting for V_0 from above gives

$$R^3 = \frac{\pi R^2 L}{2\pi} \Rightarrow L = 2R$$

Thus, the surface area of the cylinder will be minimized when the cylinder volume is fixed and the cylinder length is equal to the cylinder diameter. (Minimizing surface area for a fixed length is left as an exercise for the student.)

4.2.2 Torsion Revisited

We now want to study how nature uses shape as an integral part of structural design. First, recall from Chapter 3 that for the torsion of homogeneous linear elastic circular shafts, the only nonzero stress and strain components are related by

$$\tau_{x\theta} = G\gamma_{x\theta} \tag{4.6}$$

where G is the *shear modulus* of the material. We can now substitute the constitutive relation (4.6) into the equilibrium relation (3.6), using the shear strain–displacement relation (3.11b):

$$T = \int_A r(G\gamma_{x\theta})\,dA = G\int_A r\left(r\frac{d\phi}{dx}\right)dA \tag{4.7}$$

For constant twist gradient, we have the *moment–twist (gradient)* relation

$$T = GJ\frac{d\phi}{dx} \tag{4.8}$$

where

$$J = \int_A r^2\,dA \tag{4.9}$$

is the *polar moment of inertia* of the cross section about the cylinder axis.

For a circular cross section of diameter $D = 2R$, the integration indicated in equation (4.9) is easily performed since $dA = r\,dr\,d\theta$:

$$J = \int_0^{2\pi}\int_0^R r^2 r\,dr\,d\theta = 2\pi\frac{R^4}{4} = \pi\frac{R^4}{2} = \pi\frac{D^4}{32} \tag{4.10}$$

We can recast the moment–twist relation (4.8) in terms of the twist gradient

$$\frac{d\phi}{dx} = \frac{T}{GJ} \tag{4.11}$$

Then the total angle of twist, for a prismatic shaft of length L under constant torque of magnitude T, is found from integration of equation (4.11):

$$\phi_{max} = \int_0^L \frac{T}{GJ} dx = \frac{TL}{GJ} \quad (4.12)$$

where ϕ is in radians. (If the upper limit of integration was some intermediate point x, then the integration would supply the twist angle at x or $\phi(x)$.)

Recasting equation (4.12) in terms of applied torque gives

$$T = K\phi \quad (4.13)$$

Notice the similarity in form to the axial equation $F = ku$. The term K on the right-hand side of equation (4.13) is the *torsional stiffness* of the shaft

$$K = \frac{GJ}{L} \quad (4.14)$$

(Compare K to $k = EA/L$.) K gives the applied torque per unit twist angle ((N m)/rad).

Finally, we can substitute the shear strain–displacement relation (3.11b) into the constitutive relation (4.6) to get the shear stress in terms of the applied torque:

$$\tau_{x\theta} = Gr\frac{d\phi}{dx} = Gr\frac{T}{GJ} = \frac{Tr}{J} \quad (4.15)$$

(We will see later that there is an analogous expression for the flexural stress in a beam in terms of the applied bending moment.) By inspection of equation (4.15), we get

$$\gamma_{x\theta} = \frac{rT}{GJ} = \frac{rT}{KL} \quad (4.16)$$

Equation (4.16) shows that the shear strain is inversely proportional to the polar moment of inertia and torsional stiffness: As the stiffness decreases, the shear strain increases.

Example 4.6

A 2-ft-long steel shaft of 1¼ in. diameter is subject to a torque of 100 ft lb. Determine the maximum shear stress in the shaft and the total angle of twist.

$$R = \frac{1.25}{2} \text{ in.}, \quad L = 24 \text{ in.}, \quad G = 11.5 \times 10^6 \frac{\text{lb}}{\text{in.}^2}, \quad T = 100 \text{ ft lb}$$

$$J = \frac{\pi R^4}{2}$$
$$= 0.24 \text{ in.}^4$$

$$\tau_{max} = \frac{TR}{J}$$
$$= 3.129 \times 10^3 \frac{\text{lb}}{\text{in.}^2}$$

$$\phi_{\max} = \frac{TL}{GJ}$$
$$= 0.01 \text{ rad}$$
$$= 0.599 \text{ deg}$$

Example 4.7

A solid tapered shaft has a diameter that varies linearly from D_0 at $x = 0$ to $2D_0$ at $x = L$ as shown in Figure E.4.3a. It is subjected to an end torque T at $x = 0$ and is attached to a rigid wall at $x = L$. The shear modulus of the shaft is G.

  a. Determine an expression for the maximum (cross-sectional) shear stress in the tapered shaft as a function of the distance x from the origin of coordinates.
  b. Determine an expression for the total angle of twist of the shaft.

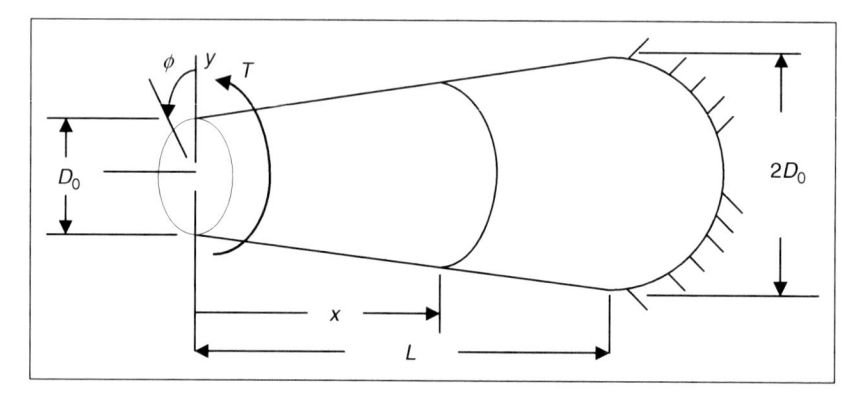

Figure E.4.3a Tapered shaft definition sketch.

Consider a cutting plane coincident with the $z = 0$ plane. The outer fiber exposed by this plane has an equation:

$$y(x) = R(x) = R_0\left(1 + c\frac{x}{L}\right)$$

where $R_0 = D_0/2$ and $c = 1$ for the problem at hand. Then

$$J(x) = \frac{\pi R^4(x)}{2} = \frac{\pi}{2} R_0^4 \left(1 + \frac{x}{L}\right)^4$$

Now with $T(x) = T$

$$\tau_{x\theta,\max} = \frac{T(x)R(x)}{J(x)} = \frac{2T}{\pi R_0^3(1 + (x/L))^3}$$

To understand what this result means, let's plot it:

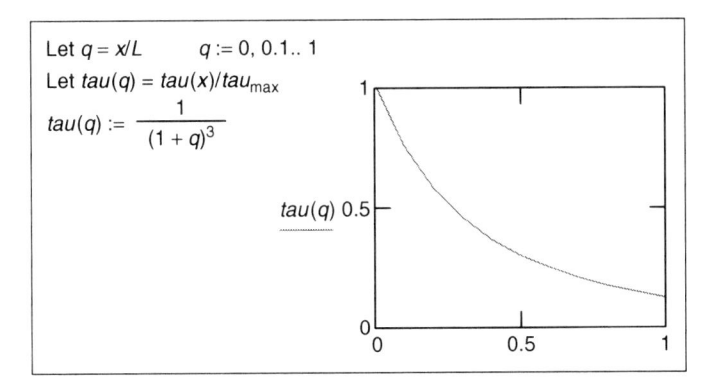

```
Let q = x/L      q := 0, 0.1.. 1
Let tau(q) = tau(x)/tau_max
tau(q) := 1/(1 + q)^3
```

Figure E.4.3b Non-dimensional shear stress vs. shaft length

We can see that the maximum shear stress exists at the small end ($x = 0$) and decays to ⅛ of that value at the large end ($x = L$). This makes sense since the smallest torsion stiffness $K(x)$ exists at the smallest end, hence the strain and thus stress are highest there.

The total angle of twist is found from

$$\phi(L) = \int_0^L \frac{T}{GJ(x)} dx$$

(Note that if the upper limit were "x" instead of "L," we would then find the twist deflection as a function of x along the shaft.) Substituting for $J(x)$ and performing the indicated integration gives

$$\phi(L) = \frac{2T}{\pi R_0^4 G} \int_0^L \frac{dx}{(1 + (x/L))^4} = \cdots = \frac{-2TL}{3\pi R_0^4 G} \left[\frac{1}{(1 + (x/L))^3} \right]_0^L = \cdots = \frac{7TL}{8\pi R_0^4 G}$$

(The intermediate steps missing above are left as an exercise for the student. Problems associated with this example are given at the end of the chapter.)

4.2.3 Thin-Walled Torsion Structures

Torsion structures needn't only be of solid section, they may be hollow as well. A hollow circular cross section would have inner radius R_i, outer radius R_o, wall thickness $h = R_o - R_i$, and average radius R. If the wall thickness h is very small compared to R, the shear stress is approximately constant across the wall thickness. That is,

$$\tau h = \text{constant} \tag{4.17}$$

The quantity τh is called the *shear flow* (see Jenkins and Khanna, 2005). Note that this implies that the largest shear stress occurs where the wall thickness is smallest. (Hereafter, we drop the subscripts on τ and ϕ to avoid any confusion.)

We now need to relate the shear flow to the applied torque T. An increment of torque dT is applied by the shear flow as

$$dT = r\tau h\, ds$$

where ds is an increment of arc length along the mean circumference l_m. Integrating along the entire mean length l_m gives:

$$T = \tau h \int_0^{l_m} r\, ds = 2\tau h A_m \quad (4.18)$$

where A_m is the mean area enclosed by the mean circumference.

Finally, we then have a useful relationship for the shear stress on a thin-walled shaft of arbitrary cross section:

$$\tau = \frac{T}{2 h A_m} \quad (4.19)$$

4.2.4 Thin-Walled Shafts of Closed Cross Section

Consider a hollow plant stem of thin circular closed cross section as shown in Figure 4.6.

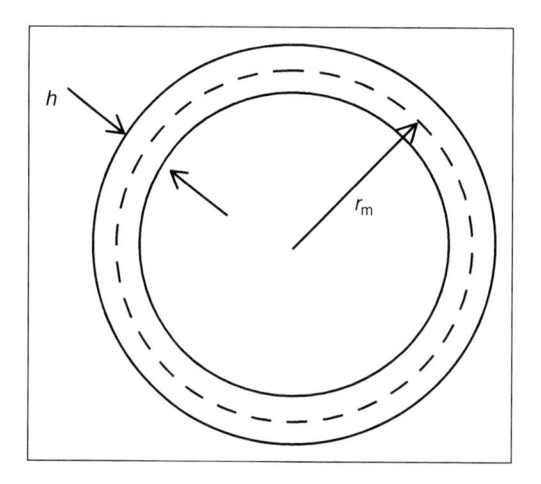

Figure 4.6. Plant stem of thin circular closed cross section. The mean radius is given by r_m, and h is the wall thickness.

The polar moment of inertia J is calculated in the usual way as before:

$$J = \frac{\pi}{2}\left[\left(r_m + \frac{h}{2}\right)^4 - \left(r_m - \frac{h}{2}\right)^4\right] \quad (4.20)$$

After expansion and some simplification (left as an exercise for the student), this becomes

$$J = \frac{\pi r_m h}{2}(4 r_m^2 + h^2) = 2\pi r_m^3 h\left[1 + \left(\frac{h}{2 r_m}\right)^2\right] \quad (4.21)$$

Figure 2.9. Mangrove roots. Roots play both structural and mass transport roles and are structured accordingly (HD). Redundant load paths (roots) allow for changing conditions (JGE). The roots live in an environment that would challenge many engineered structures.

http://commons.wikimedia.org/wiki/File:Mangrove.jpg.

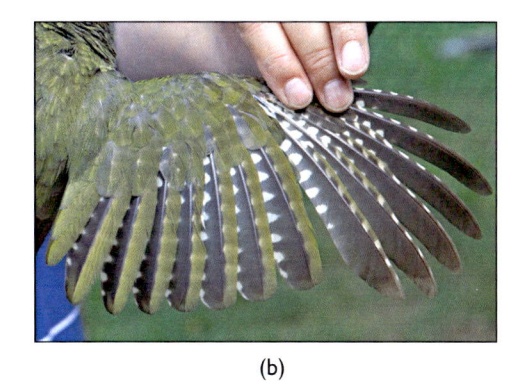

(a) (b)

Figure 2.10a–b Wings. Different elements support tensile loads, compressive loads, or control airflow (PSF). (a) Flexibility gives maneuverability and ability to change shape as needed (LM). http://en.wikipedia.org/wiki/File:Big-eared-townsend-fledermaus.jpg. (b) Many feathers give damage tolerance and can be replaced individually (PSF).

http://commons.wikimedia.org/wiki/File:Green_Woodpecker_wing.jpg.

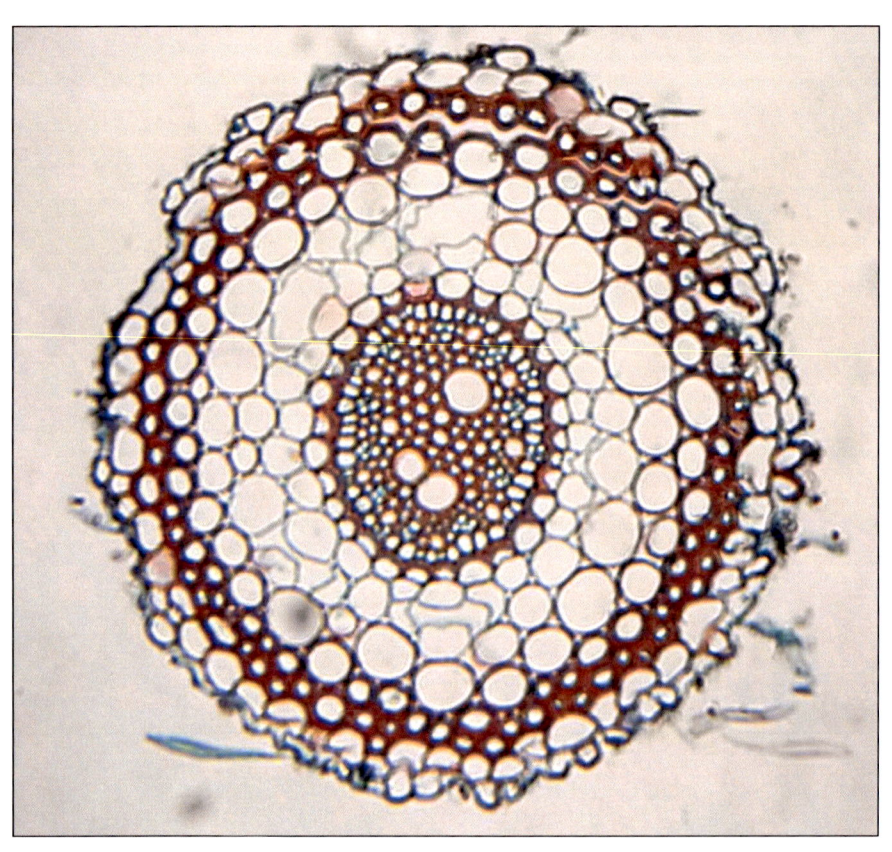

Figure 4.13. Material structural hierarchy in the root of the tritium plant.
http://commons.wikimedia.org/wiki/File:TritiumRootCross.png.

Figure 5.4. Fry of the paradise fish or paradise gouramis (*Macropodus opercularis*) in their bubble nest. http://commons.wikimedia.org/wiki/File:Spawn_hatched_fish_2.JPG.

Figure 5.6. The bullfrog's throat is an inflatable membrane.
http://en.wikipedia.org/wiki/File:Bullfrog_-_natures_pics.jpg.

Figure 5.9. Scientific high-altitude balloon.
http://commons.wikimedia.org/wiki/File:Wallops_Balloon_With_BESS_Payload_DSC00088.JPG.

Figure 5.10. The 20 m ATK solar sail undergoing deployment testing in the NASA Plum Brook vacuum chamber. http://commons.wikimedia.org/wiki/File:Solar_sail_tests.jpg.

Figure 6.4. Sap infusing into a tree's wound. http://commons.wikimedia.org/wiki/File:Tree_sap.jpg.

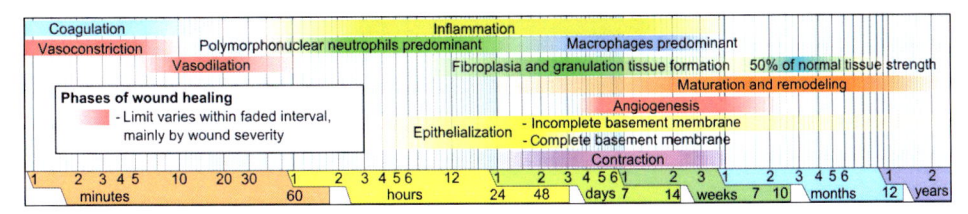

Figure 6.5. Phases of wound healing. http://en.wikipedia.org/wiki/File:Wound_healing_phases.png.

Figure E.7.1. Three-horned rhinoceros beetle. http://commons.wikimedia.org/wiki/File:Rhinoceros_beetle.jpg.

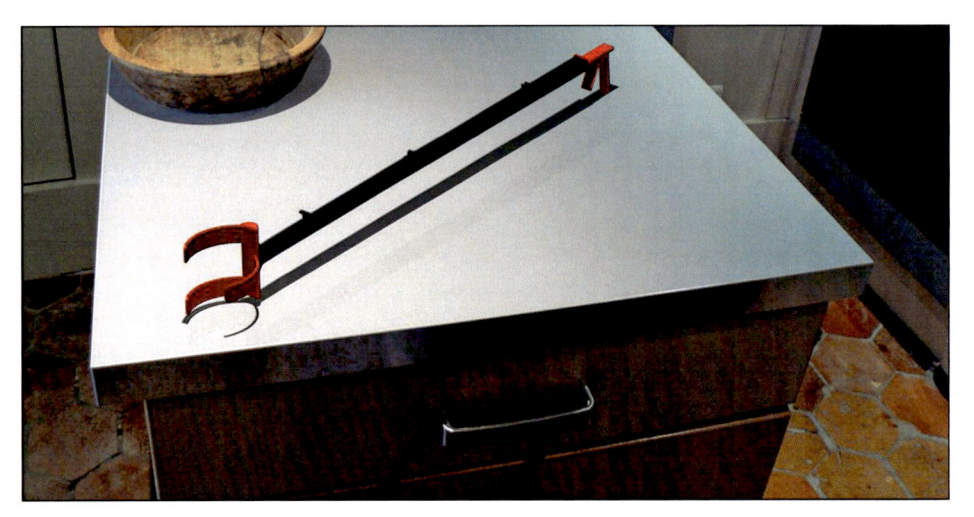

Figure E.7.2. Concept of possible new product inspired by the three-horned rhinoceros beetle.

Now for a thin wall, where $h/r_m \ll 1$ and $(h/2r_m)^2 \ll 1$, equation (4.21) simplifies to

$$J \approx 2\pi r_m^3 h = 2A_m r_m h \tag{4.22}$$

where A_m is the area enclosed by the mean radius.

The shear stress is then

$$\tau = \frac{Tr_m}{J} \approx \frac{Tr_m}{2\pi r_m^3 h} = \frac{T}{2\pi r_m^2 h} = \frac{T}{2A_m h} \tag{4.23}$$

For $h/r_m = 0.2, 0.1$, and 0.05, the approximate shear stress values are 92%, 95%, and 98% of the exact shear stress, respectively. Note also that since τ is assumed not to vary across the thickness, the stress found above is the maximum shear stress. The total twist angle is given as before:

$$\phi_{\text{total}} = \frac{TL}{GJ} = \frac{TL}{2GA_m r_m h} \tag{4.24}$$

Of course, for the thin-walled closed circular cross section, the approximation to J provides us little advantage. However, we can apply this same approximation technique to more complex thin-walled shapes, of either open or closed section, where calculating the exact polar moments of inertia would be difficult. (In these cases of noncircular cross section, the factor J is no longer the polar moment of inertia but is called more generally the *torsion constant*, and is in general less than the polar moment of inertia.)

A formal method for finding the torsion constant can be developed from strain energy considerations (Jenkins and Khanna, 2005). In that case, J is the torsion constant found from

$$J = \frac{4A_m^2}{\displaystyle\int_0^{l_m} (ds/h)} \tag{4.25}$$

Example 4.8

Consider, for example, the thin square closed cross section shown in Figure E.4.4.

The above formula for J can be applied here. The value of the integral in the denominator is simply $4b/h$, A_m is b^2, giving $J = b^3 h$. Hence, the average shear stress (maximum shear stress must take into account the stress concentrations in the corners) is

$$\tau = \frac{T}{2hA_m} = \frac{T}{2hb^2}$$

and the total twist angle is

$$\phi_{\text{total}} = \frac{TL}{Gb^3 h}$$

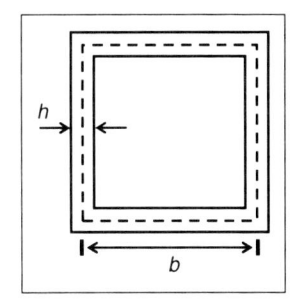

Figure E.4.4. Shaft of thin square closed cross section. A side has a mean length given by b and a wall thickness h.

4.2.5 Thin-Walled Shafts of Open Cross Section

Slicing a thin-walled shaft of closed circular cross section longitudinally results in an *open cross section* (Figure 4.7).

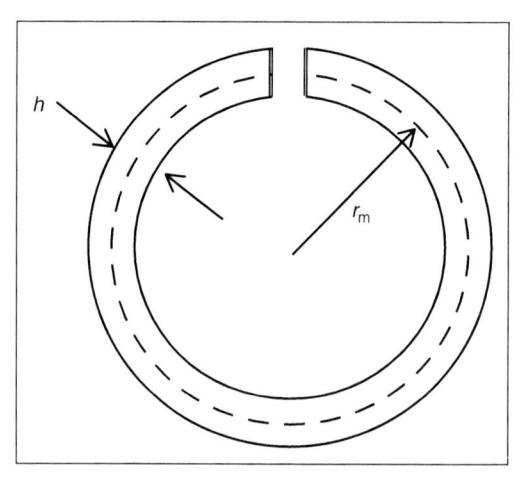

Figure 4.7. A longitudinal cut transforms a closed cross section into an open cross section.

However, this cut "releases" the shear stress acting longitudinally along the shaft. This significantly increases flexibility of the structure in torsion, and hence thin-walled open sections are considerably less efficient in torsion than comparable closed cross sections.

The exact form of the torsion constant for a thin solid rectangular section b by h ($b > h$) (Figure 4.8) can be shown to be (Young, 1989)

$$J = \frac{bh^3}{16}\left[\frac{16}{3} - 3.36\frac{h}{b}\left(1 - \frac{h^4}{12b^4}\right)\right] \qquad (4.26)$$

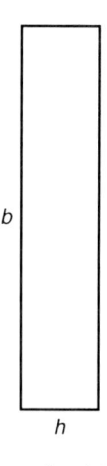

Figure 4.8. Cross section of a thin solid rectangular shaft.

For $h/b \ll 1$, $J \approx bh^3/3$. Thus, the torsion constant for open sections can be approximated as a sum of thin rectangular sections. Figure 4.9 provides several examples.

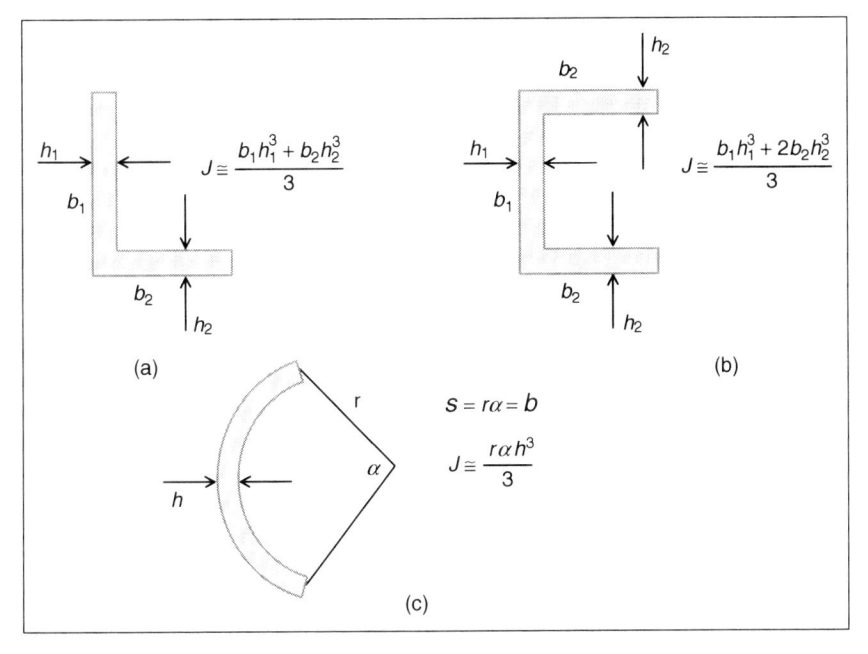

Figure 4.9. Approximate torsion constants for some thin-walled open sections: (a) angle, (b) channel, and (c) circular arc.

The (average) shear stress (away from sharp corners) and the total twist angle are found as

$$\tau_{\text{avg}} = \frac{Th}{J}, \quad \phi = \frac{TL}{GJ} \tag{4.27}$$

4.2.6 Torsion Shape Factor

We can now begin to see how we might characterize the shape efficiency of a given cross-sectional configuration. We define a *shape factor* β that compares the relative efficiency of a given cross section to a reference cross section. Following Ashby (2005), we arbitrarily take a solid circular shaft as the torsion reference. This reference shaft has length L_0, cross-sectional area $A_0 = pR_0^2$, shear modulus G_0, and torsion constant

$$J_0 = \frac{\pi}{2}R_0^4 = \frac{1}{2}A_0 R_0^2 \tag{4.28}$$

Under the action of a torque T, the reference shaft has a maximum shear stress and total twist angle given as before:

$$\tau_0 = \frac{TR_0}{J_0} = \frac{2T}{A_0 R_0} \tag{4.29}$$

Chapter 4

$$\phi_0 = \frac{TL_0}{G_0 J_0} = \frac{2TL_0}{G_0 A_0 R_0^2} \qquad (4.30)$$

Then we use β to compare the efficiency of other shaft configurations having shear modulus G, but (departing from Ashby) <u>equivalent cross-sectional area A_0 and length $L = L_0$</u>, under the action of the same torque T. That is,

$$\beta_\tau = \frac{\tau}{\tau_0} = \frac{\tau}{2T/A_0 R_0} \qquad (4.31)$$

$$\beta_\phi = \frac{\phi}{\phi_0} = \frac{TL/GJ}{TL_0/G_0 J_0} = \frac{G_0 J_0}{GJ} = \frac{G_0}{G} \frac{A_0 R_0^2}{2J} \qquad (4.32)$$

A value of β less than 1 means that the shaft under consideration is more efficient than the reference solid circular shaft, that is, for the same load and weight, the stress and/or stiffness are less than that for the reference shaft.

Example 4.9

Determine β for the thin-walled circular shaft of radius R and wall thickness h ($h \ll R$).

$$A_c = 2\pi r_m h = A_0 = \pi R_0^2 \Rightarrow R_0 = \sqrt{2 r_m h}$$

$$A_{mc} = \pi r_m^2$$

$$\beta_\tau = \frac{\tau_c}{\tau_0} = \frac{T/2hA_{mc}}{2T/A_0 R_0} = \frac{A_0 R_0}{4hA_{mc}} = \frac{(2\pi r_m h)\sqrt{2 r_m h}}{4h\pi r_m^2} = \cdots = \sqrt{\frac{h}{2r_m}} = 0.707\sqrt{\frac{h}{r_m}}$$

$$\beta_\phi = \frac{J_0}{J_c} = \frac{(1/2)A_0 R_0^2}{2\pi r_m^3 h} = \frac{1}{4\pi} \frac{(2\pi r_m h)(2 r_m h)}{r_m^3 h} = \frac{h}{r_m}$$

Table E.4.2 shows a comparison of both the strength and stiffness efficiency of the thin-walled circular shaft for various wall thicknesses (compared to the solid circular shaft of same cross-sectional area and load). As shown, as the wall thickness becomes thinner, the shaft becomes more efficient. (Note that this trend cannot go on indefinitely, since the shaft is getting larger to keep the area constant and equal to A_0. As the wall thickness gets thinner, other failure modes will start to take place, such as wall buckling.)

Table E.4.2

$\dfrac{h}{r_m}$	β_τ	β_ϕ
1/10	0.224	0.1
1/100	0.0707	0.01
1/1,000	0.0224	0.001

Example 4.10

Determine the torsional efficiency of a thin-walled tube of open circular cross section.

For the section shown in Figure 4.8, we find the torsion constant from Figure 4.10 as (with $\alpha = 2\pi$)

$$J_u = \frac{2\pi r_m h^3}{3}$$

Enforcing equal weight means equal cross-sectional areas between the solid (A_0) and open (A_u) circular sections or

$$A_u = 2\pi r_m h = A_0 = \pi R_0^2 \Rightarrow R_0 = \sqrt{2 r_m h}$$

Then

$$\beta_\phi = \frac{J_0}{J_u} = \frac{2\pi r_m^2 h^2}{2\pi r_m h^3/3} = 3\frac{r_m}{h}$$

and

$$\beta_\tau = \frac{\tau_u}{\tau_0} = \frac{Th/J_u}{TR_0/J_0} = \frac{h}{R_0}\beta_\phi = \frac{h}{\sqrt{2 r_m h}}\frac{3 r_m}{h} = 3\sqrt{\frac{r_m^2 h^2}{2 r_m h}} = \frac{3}{\sqrt{2}}\sqrt{\frac{r_m}{h}}$$

Table E.4.3 shows a comparison of both the strength and stiffness efficiency of the thin-walled open circular shaft for various wall thicknesses (compared to the solid circular shaft of same cross-sectional area and load). It is clear that the open section is considerably inefficient compared to the reference section. As shown, as the wall thickness becomes thinner, the shaft becomes more inefficient. (Note that this trend cannot go on indefinitely, since the shaft is getting larger to keep the area constant and equal to A_0. As the wall thickness gets thinner, other failure modes will start to take place, such as wall buckling.)

Table E.4.3

$\frac{h}{r_m}$	β_τ	β_ϕ
1/10	30	6.7
1/100	300	12.1
1/1,000	3,000	67.1

4.2.7 Bending Shape Factors

For a reference beam of solid *square* cross section of side b and length L, the maximum bending stress is given by (Jenkins and Khanna, 2005)

$$\sigma_s = \frac{M_{max}(b/c)}{b^4/12} = \frac{6M_{max}}{b^3} \tag{4.33}$$

Chapter 4

Beam deflection follows a form

$$\text{Displacement} \sim \frac{\text{Load}}{\text{Stiffness}}$$

Boundary conditions and the type of loading greatly determine the exact form of the response. However, a large majority of beams follow

$$\delta \sim \frac{\text{Load}}{EI/L^3}$$

where δ is the maximum lateral displacement.

We define a bending stress shape factor as

$$\beta_\sigma = \frac{\sigma}{\sigma_s} \tag{4.34}$$

where the reference section is the solid square section of side b. If we take the same reference beam as for $\beta\sigma$, $I_s = A_s^2/12$, $A_s = b^2$, and the deflection shape factor is

$$\beta_\delta = \frac{\delta}{\delta_s}$$
$$= \frac{E_s I_s}{EI}$$
$$= \frac{E_s b^4}{12 EI} \tag{4.35}$$

Example 4.11

As we did above, we take the square cross section of side b as the reference shape. Then $A_s = b^2$, $I_s = b^4/12 = A_s^2/12$, and $c_s = b/2$. The maximum stress for this cross section is then

$$\sigma_s = \frac{M_{\max}(b/2)}{b^4/12} = \frac{6 M_{\max}}{b^3}$$

Let's compare the reference square section to a circular section of radius R but equivalent area A_s (see Figure E.4.5).

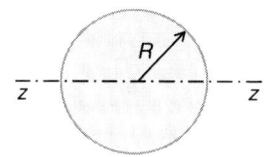

Figure E.4.5. Circular beam cross section.

Then

$$A_0 = \pi R_0^2 = A_s = b^2 \Rightarrow R_0 = \frac{b}{\sqrt{\pi}}$$

$$I_{zz} = I_0 = \frac{\pi R_0^4}{4} = \frac{b^4}{4\pi}$$

76

and $c = R_0$. Finally,

$$\beta_\sigma = \frac{\sigma_0}{\sigma_s} = \frac{(M_{max} C_0)/I_0}{6M_{max}/b^3} = \frac{(b/\sqrt{\pi})(4\pi/b^4)}{6/b^3} = \frac{2\pi}{3\sqrt{\pi}} = 1.18$$

$$\beta_\delta = \frac{\delta_0}{\delta_s} = \frac{I_s}{I_0} = \frac{b^4/12}{b^4/4\pi} = \frac{4\pi}{3} = 1.05$$

This result says that the circular cross section is not as efficient in bending as the square cross section of the same area under the same load, by approximately 18% in stress and 5% in deformation. This makes sense if you think in terms of the amount of load-carrying fibers available where the stress is highest.

4.2.8 Combined Shape Factor

We've looked at torsion and bending shape performance separately. What if we want to optimize performance in both torsion and flexure? We would consider a combined shape factor that is composed of torsion shape factor and bending shape factor. (Ashby (2005) introduces a microstructural shape factor and combines it with the "macrostructural" shape factor for a different kind of combined shape factor.) Further exploration of combined shape factors is left as a homework problem for the student.

4.3 NATURAL MATERIALS

Nature has a limited selection of materials to work with, especially compared to the vast universe of engineering materials. There is nothing particularly special about the "building blocks" of natural materials—it is how they are combined that gives these materials such exceptional properties. In the brief overview that follows, we will focus on the subset of natural materials that comprise biological materials. For example, we will not discuss the various minerals found in geology.

General features of all biological materials include the following features (in no particular order):

1. Growth
2. Self-repair
3. Aging
4. Adaptation
5. Multifunctionality
6. Hierarchy

Of these characteristics, the ability to grow new tissue, bone, etc., is a hallmark of biological organisms and one that will not likely be realized in engineered materials. All

Chapter 4

of the other features may be emulated to some extent in engineering. The overarching characteristic of structural hierarchy deserves further discussion.

4.3.1 Structural Hierarchy

Biological materials contain levels of organization that build upon one another, resulting in their unique properties (Myers et al., 2011). To understand structural hierarchy in biological materials, first consider the organizational hierarchy shown in Figure 4.10. The mapping shown in Figure 4.10 is a common organizational paradigm in business and government. In this hierarchical organization, the enterprise is finely subdivided into many specialized units. There are well-defined chains of command wherein each unit is singularly subordinate to another unit, except that all are subordinate to the unit at the top of the hierarchy.

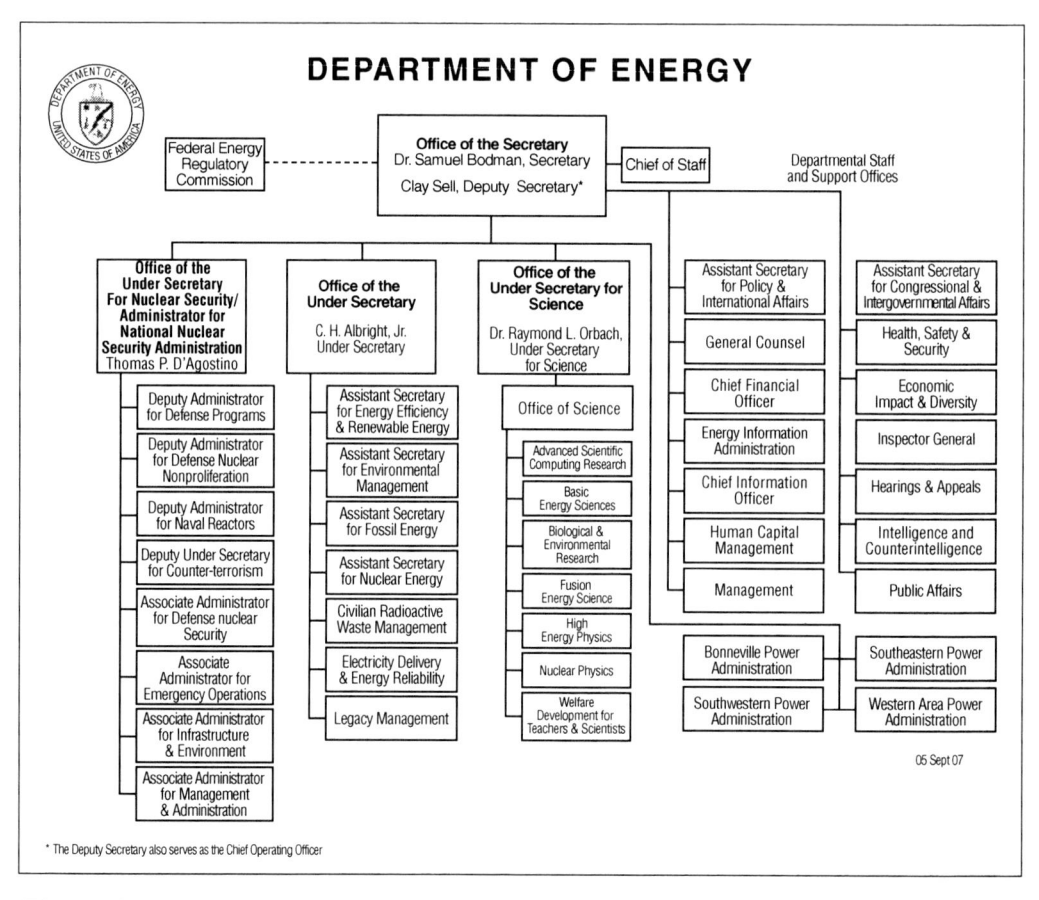

Figure 4.10. The hierarchical organization. http://commons.wikimedia.org/wiki/File:DOE_org_crart.PNG.

As another example, in biology as a whole a hierarchy can be identified and might go something like Figure 4.11.

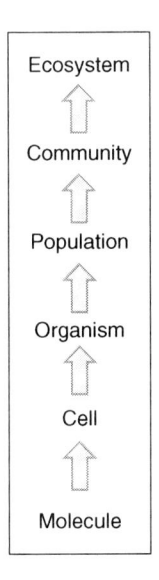

```
Ecosystem
   ⇧
Community
   ⇧
Population
   ⇧
Organism
   ⇧
  Cell
   ⇧
Molecule
```

Figure 4.11. A hierarchy in systems biology.

The hierarchical organization of biological material constituents results in a synergism evidenced by the high performance of these materials. Hierarchical material characteristics include

1. recurrent use of a relatively few molecular constituents such that widely varying global properties result;
2. controlled orientation;
3. interfaces that provide opportunities for DDOF;
4. fatigue and fracture resistance;
5. complex shapes.

Traditional engineered materials, such as steel or concrete, exhibit very little hierarchy. On an imagined "walk" through an engineering structure, every stop would look about the same as every other. But in even simple biological structures, such as the plant root (Figure 4.12), the material diversity is high. For example, in the human tendon, the smallest building blocks are collagen molecules (see below) of approximately 1.5 nm length. These in turn form microfibrils (a few nanometers in length), then subfibrils and fibrils (50–500 nm). The fibrils form fascicles (50–300 μm), which finally form the tendon.

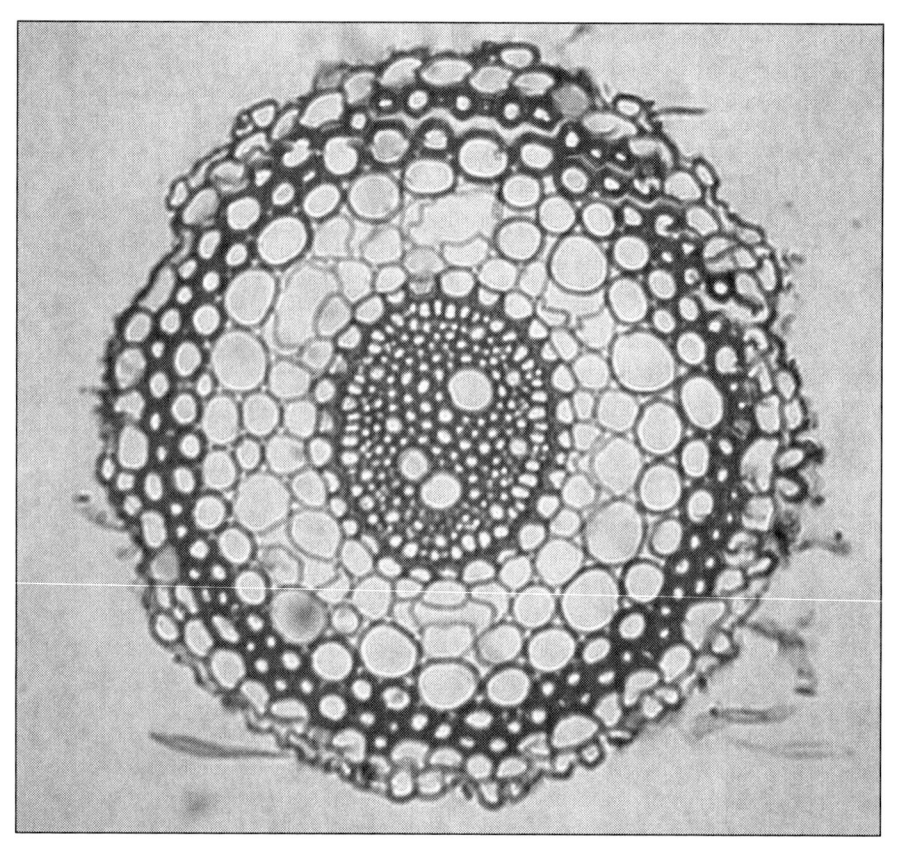

Figure 4.12. Material structural hierarchy in the root of the tritium plant.
http://commons.wikimedia.org/wiki/File:TritiumRootCross.png.

4.3.2 Biological Material Building Blocks

Basic biological structural material groups include natural

1. ceramics—calcium carbonate, calcium phosphate (stiff materials);
2. polymers—proteins (collagen, keratin, silk), polysaccharides (cellulose, chitin);
3. elastomers—elastin, reslin, abductin;
4. composites—wood, nacre.

Basic biological material constituents ("building blocks") include the following:

1. Proteins—proteins are organic compounds made of amino acids; they are widely used as enzymes for catalysts, for signal transmission, and for structural materials. Structural proteins may be fibrous, e.g., collagen, elastin, keratin, and actin.
2. Collagen—most abundant protein in mammals, 25–35% of total protein, main load-carrying element in tendons, cartilage, bone, and blood vessels.

3. Keratin—protective covering of all land vertebrates, e.g., hair, skin, nails, and hooves.
4. Actin—used in muscle tissue.
5. Chitin—carbohydrate, derivative of glucose, second most abundant fibrous material (next to collagen), used, e.g., in insect exoskeletons.

Tables 4.1 and 4.2 provide data on some basic biological material mechanical properties and relative distribution in humans, respectively. (Additional material property data can be found in Appendix 5.)

Table 4.1. Mechanical properties of some biological materials (steel shown for comparison only)

	Young's Modulus (MPa)	Tensile Strength (MPa)
Elastin	0.6	2–20
Collagen	10^3	80–100
Bone	10^4	100
Silk	10^4	1,000
Chitin	10^4	20–1,000
Mild steel	2×10^5	500

Table 4.2. Relative distribution of biological materials in humans (after Myer et al., 2010)

	Weight Percentage in Human Body
Proteins	17
Lipids	15
Carbohydrates	1
Minerals	7
DNA, RNA	2
Water	58

There is much more to say about biological materials (see, e.g., Vincent, 1990), but we have to end our study here. Before we conclude, however, we should note that perhaps the most important function of biological materials is to replicate cellular information. This is part of what has been called the *central dogma of biology*, that is, "DNA makes RNA makes protein."

KEY POINTS TO REMEMBER

- Size scaling is very important in biology and can drive evolutionary processes.
- Shape also plays an integral role in natural structures.
- The concept of a shape factor is a powerful tool for comparing structural efficiencies of various shapes.
- Structural hierarchy is one of several important features of natural materials.
- Natural materials achieve high performance by innovative use of simple material building blocks.

REFERENCES

Ashby, M. F. 2005. *Materials Selection in Mechanical Design*, 3rd ed. Elsevier.

Calder, W. A. 1984. *Size, Function, and Life History*. Harvard University Press.

Jenkins, C. H., and S. K. Khanna. 2005. *Mechanics of Materials: A Modern Integration of Mechanics and Materials in Structural Design*. Elsevier.

Mandelbrot, B. B. 1983. *The Fractal Geometry of Nature*. San Francisco: W.H. Freeman.

Morey-Holton, E. R. 2003. "The Impact of Gravity on Life." In *Evolution on Planet Earth: the Impact of the Physical Environment*, edited by Lynn J. Rothschild and Adrian Lister. Academic Press.

Myers, M. A., P.-Y. Chen, M. I. Lopez, Y. Seki, and A. Y. M. Lin. 2011. "Biological Materials: A Materials Science Approach." *J. Mech. Behav. Biomed. Mater.* 4:626–657.

Niklas, K. J. 1992. *Plant Biomechanics*. University of Chicago Press.

Thompson, D. W. 1992. *On Growth and Form*. Dover.

Vincent, J. F. V. 1990. *Structural Biomaterials*. Princeton University Press.

PROBLEMS

4.1. Allometry of trees

    a. Make a good estimate of the height and measure the trunk diameter of at least three trees, plot the results, and determine the $D \sim L$ scaling. For the best data, the trees should be of the same species (e.g., spruce) and different heights and should have a well-defined trunk (e.g., spruce).

    b. Discuss any differences between the Greenhill and Euler buckling models.

4.2. What can be learned from the allometry of bird eggs (see example in the text) and how might that provide engineering inspiration?

4.3. Carefully sketch the first four iterations of the Koch snowflake.

4.4. For the branching example worked in class,
   a. what is the change in A_x/V if we consider the same flow continuity but with $L_2 = 2L_1 = L_0$?
   b. what is the change in A_x/V if we consider the same flow continuity but with a second iteration of symmetric branching and all leg lengths equal?

4.5. For a right circular closed cylinder of radius R and length L,
   a. for a fixed volume, minimize the surface area with respect to L,
   b. find the optimal design that minimizes the interior volume for a fixed surface area.

4.6. A hollow tapered shaft has an outer diameter that varies linearly from D_0 at $x = 0$ to $2D_0$ at $x = L$. The inner diameter is a constant $D_i = D_0/2$. It is subjected to an end torque T_0 at $x = 0$ and is attached to a rigid wall at $x = L$. The shear modulus of the shaft is G.
   a. Determine an expression for the maximum (cross-sectional) shear stress in the tapered shaft as a function of the distance x from the origin of coordinates.
   b. Determine an expression for the total angle of twist of the shaft.

4.7. Compute the torsion and bending shape factors β for the thin-walled square shaft of mean side length a and wall thickness h. Compare the shape efficiency for values of $h/a = 0.1$, 0.01, and 0.001.

4.8. Compute the torsion shape factors β for the thin-walled "u" channel.

4.9. Compute the deflection (torsion and bending) shape factors β for the "half-round" shape that is a first approximation to a petiole shape (see Chapter 5).

4.10. Combined shape factors
   a. For the following shapes, compare three combined shape factors as shown in the table.

Shape	$\beta_\phi \beta_\delta$	$(\beta_\phi \beta_\delta)^{1/2}$	$(\beta_\phi^2 + \beta_\delta^2)^{1/2}$
Thin-walled closed circular cylinder			
Thin-walled open circular cylinder			
Half-round			
(see part b)			

   b. Recompute the bending deflection shape factor for the half-round shape using the solid circular cylinder as the reference shape, and then recompute the combined shape factors and add the results to the fourth row of the table.

4.11. A pressurized cylindrical membrane has a profile given by

$$R(x) = (R_0 - R_{max})\left(\frac{2x}{L} - 1\right)^2 + R_{max}$$

a. Determine the total twist angle using $R(x)$ above and $R_{max} = cR_0$, where c is a constant.
b. If the cylinder deforms like a beam with fixed supports under a uniform load (thus a "sheet" rather than a true membrane), the profile would go like:

$$R(x) = 16(R_{max} - R_0)x^2(1-x)^2 + R_0$$

Determine the total twist angle using $R(x)$ above and $R_{max} = cR_0$, where c is a constant.

4.12. A long cylinder 50.8 mm in diameter and made from 25-μm-thick polyethylene is placed under an internal pressure of 100 Pa. Using the Standard Linear Solid model, what is the cylinder diameter after 8 h?

4.13. A strip of pig skin is loaded in tension. If the skin can be represented by a Mooney–Rivlin nonlinear elastic model with coefficients $C_1 = 0.261$ MPa and $C_2 = 0.3668$ MPa, what is the tensile stress when the sample doubles its length?

CHAPTER 5

Compliant Structures

> Soft tissues are remarkably widespread in the animal world and represent a useful, safe, cheap, and efficient way of containing and organizing other materials.
>
> —*Julian F. V. Vincent*

In this chapter, we will introduce the concept of structural compliance, its dominance in natural design, and its various applications in nature and engineering. We will look briefly at the response of compliant engineering structures, in order to better understand and learn from compliance in nature.

5.1 THE COMPLIANT PARADIGM IN NATURE

Biological systems are driven to reduce energy cost in both fabrication and maintenance. The spider must make up the energy and time lost in web building as quickly as possible through prey capture. Additionally, the web must continue to perform even in the presence of significant degradation, else building begins all over again. Over eons, nature has converged on *structural compliance* or flexibility as a key component of energy efficiency. Often, the compliant solution is a less costly alternative to the stiff solution; for a given material, increasing stiffness requires additional mass in many cases. The lighter-weight compliant structure puts fewer demands on its boundaries and is often easier and faster to repair. Compliant structures are much better at carrying shock loads than their stiff counterparts. Moreover, enabling increased intelligence goes hand in hand with increased compliance—intelligence would be of little use to a rock!

A fundamental enabler of structural compliance is the *membrane*. A formal definition of *membrane* is beyond our scope, so for now let's consider a membrane as a two-dimensional surface, like the surface of a balloon (Figure 5.1). Recall from Chapter 3 that structural stiffness (or compliance) is a function of intrinsic material stiffness and structural geometry. In the case of the balloon, the membrane that forms the balloon's shape is composed of a thin elastomer. The elastic modulus of the elastomer is low and the thickness is small, with the result that the structural stiffness is small and the structural compliance is large. (If structural stiffness is k, the structural compliance is roughly $1/k$.)

Membranes are everywhere in nature. The central structural paradigm for the complex architectures of biological systems is the membrane. Hundreds of examples can be given, from the walls of cells to the bullfrog's throat (Jenkins, 2005). Lipowsky (1991) points out that the brain is composed of a complex network of membranes with a total surface area of 1,000–10,000 m^2!

Figure 5.1. A toy balloon is a quintessential membrane structure. http://www.openclipart.org/detail/105775.

A range of compliant structures exists, with the membrane representing the extremity (along with its 1D counterpart—the rope or chain). A compliant structure is any structure that, under normal service loads, exhibits large deformations. Those purposely large deformations are part of the load-carrying strategy of the structural design. Consider a sheet of paper held out in cantilever fashion (Figure 5.2a). The structure suffers large deformations in this configuration and can barely support its own weight. (The *curvature* is clearly evident in the picture, and thus a moment must be generated as an internal reaction in the paper—recall the moment–curvature relation in beam theory. Then the structure is not a membrane—membranes cannot sustain moments.) A slight change in the cantilever sheet configuration, giving it just a small amount of initial curvature, now allows the sheet to carry loads other than its own weight (Figure 5.2b).

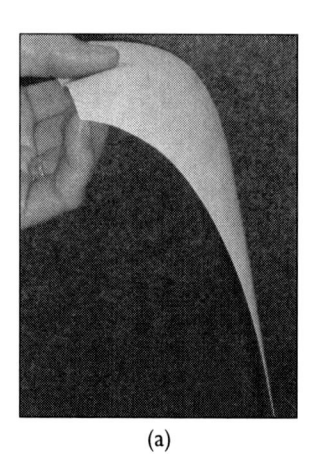

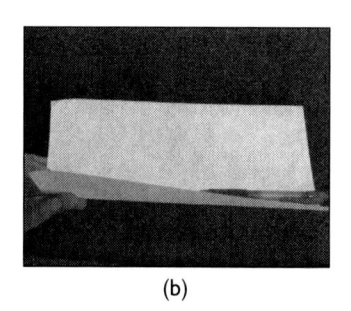

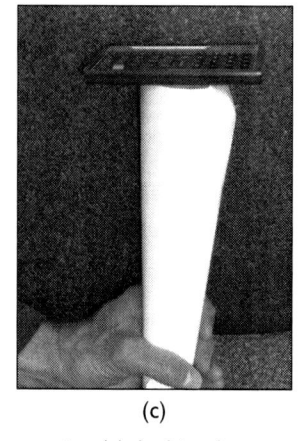

(a)   (b)   (c)

Figure 5.2. The significance of shape change on load-carrying capacity. (a) A thin sheet of paper can barely support its own weight in cantilever configuration. (b) With a slight initial curvature added, the sheet is now capable of carrying a small additional load. (c) A significant initial shape change allows the same sheet to bear a much greater load.

Finally, a wholesale shape reconfiguration allows this same sheet to carry significant loads (Figure 5.2c).

5.2 EXAMPLES OF COMPLIANT STRUCTURES IN NATURE AND ENGINEERING

In this section, several examples of compliant structures in nature and engineering are shown in Figures 5.5–5.10.

Figure 5.3. Blood cells. In this scanning electron microscope image, you can see red blood cells, several white blood cells including lymphocytes, and many small disk-shaped platelets. http://commons.wikimedia.org/wiki/File:SEM_blood_cells.jpg.

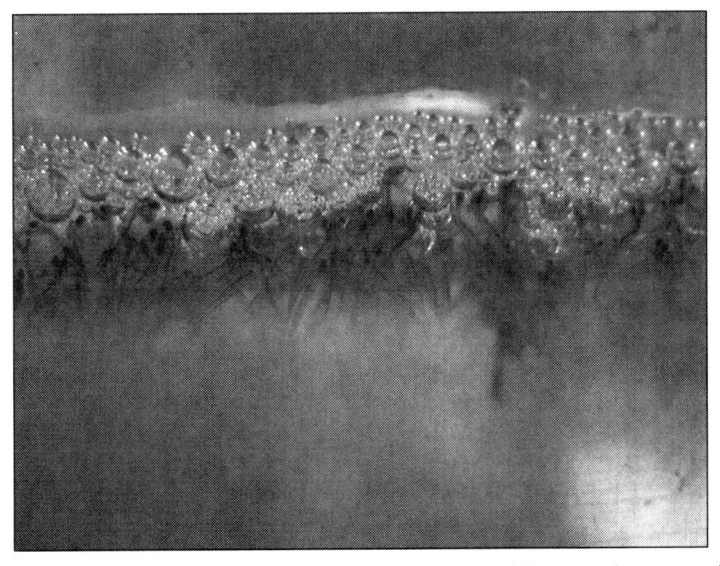

Figure 5.4. Fry of the paradise fish or paradise gouramis (*Macropodus opercularis*) in their bubble nest. http://commons.wikimedia.org/wiki/File:Spawn_hatched_fish_2.JPG.

Figure 5.5. Blades of grass are slender compliant columns.
http://commons.wikimedia.org/wiki/File:Blade_grass.jpg.

Figure 5.6. The bullfrog's throat is an inflatable membrane.
http://en.wikipedia.org/wiki/File:Bullfrog_-_natures_pics.jpg.

Figure 5.7. The City of Bridgetown's (Barbados) Grantley Adams International Airport. The translucent membrane roof provides pleasant shelter between the departure and arrival halls. http://commons.wikimedia.org/wiki/File:Sir_Grantley_Adams_Int_Airport,_Barbados-03.jpg.

Figure 5.8. A modern ram air parasail.
http://upload.wikimedia.org/wikipedia/commons/0/08/Ram_air_square.jpg.

Figure 5.9. Scientific high-altitude balloon.
http://commons.wikimedia.org/wiki/File:Wallops_Balloon_With_BESS_Payload_DSC00088.JPG.

Figure 5.10. The 20-m ATK solar sail undergoing deployment testing in the NASA Plum Brook vacuum chamber. http://commons.wikimedia.org/wiki/File:Solar_sail_tests.jpg.

5.3 COMPLIANT STRUCTURE RESPONSE ANALYSIS

Let's now take a look at models of compliant structure response. We do this in order to understand the difference between the stiff and compliant structure. We'll first look at the stiff response, which is represented by a linear model since the deformations are very small. The compliant response, with its large deformation, will require a nonlinear model. Three configurations are considered with analogs in nature:

1. Compliant columns (e.g., grass blades)
2. Compliant beams (e.g., tree branches)
3. Compliant cables (e.g., spider webs)

5.3.1 Column Buckling

Columns carry axial loads in compression and fail by buckling, either globally or locally. Stiff column response (linear response) will be considered first, followed by the buckling response of compliant columns (nonlinear response).

Stiff Column Buckling

The stiff (linear) model of column buckling follows Euler's approach (Leonard Euler, 1707–1783) and provides the *critical load* P_{cr} at which the onset of buckling occurs. Consider the cantilever column in Figure 5.11.

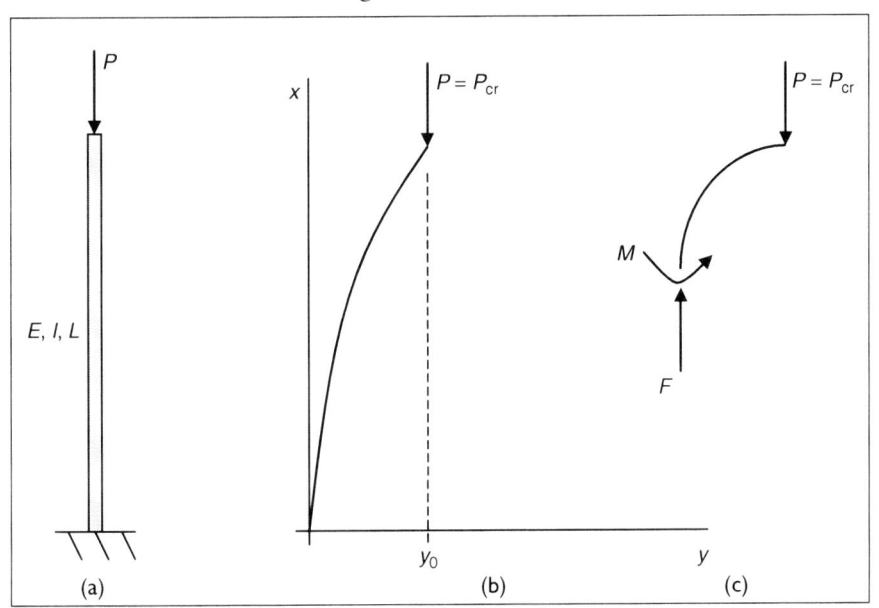

Figure 5.11. Euler column buckling: (a) initial configuration; (b) deformed configuration (deformation shown greatly exaggerated); (c) free-body diagram.

Chapter 5

The column has a length L, area moment of inertia I, and elastic modulus E, and carries axial load P. The tip deflection, y_0, at $P = P_{cr}$, is a very small quantity in the linear theory (taken to be infinitesimally small). In this case, the column curvature can be approximated by d^2y/dx^2. (The exact curvature expression is given in the next section). Then the moment–curvature relation in the standard linear beam formulation is

$$EI \frac{d^2y}{dx^2} = -M \quad (5.1)$$

where the moment at any cross section is given by

$$M = -P(y_0 - y) \quad (5.2)$$

Using equation (5.2), equation (5.1) becomes

$$\frac{d^2y}{dx^2} = -c^2(y_0 - y)$$

or

$$y'' + c^2 y = c^2 y_0 \quad (5.3)$$

where $c^2 = P/EI$ and the $'$ indicates d/dx and $''$ indicates d^2/dx^2.

The general solution of equation (5.3) is (see Problems at the end of this chapter)

$$y(x) = A \cos cx + B \sin cx + y_0 \quad (5.4)$$

where A and B are constants of integration determined from boundary conditions (BC) $y(0) = 0$, $[dy/dx]_{y=0} = 0$, then

$$y(0) = 0 = A(1) + B(0) + y_0 \Rightarrow A = -y_0$$
$$y'(0) = 0 = -Ac(0) + Bc(1) \Rightarrow B = 0$$

Equation (5.4) then becomes

$$y = y_0(1 - \cos cx) \quad (5.5)$$

The requirement that $y(L) = y_0$ gives, since $y_0 \neq 0$,

$$y(L) = y_0 + y_0 \cos cL \Rightarrow \cos cL = 0 \quad (5.6)$$

Therefore,

$$cL = (2n - 1)\frac{\pi}{2}, \quad n = 1, 2, 3, \ldots$$

Equation (5.6) determines the value of c at which buckling occurs. The smallest value of cL that satisfies equation (5.6) is the critical value and the corresponding critical load P_{cr} is then

$$cL = L\sqrt{\frac{P_{cr}}{EI}} = \frac{\pi}{2}, \quad n = 1$$

or

$$P_{cr} = \frac{\pi^2}{4} \frac{EI}{L^2} \tag{5.7}$$

Compliant Column Buckling

The critical buckling load as determined above is an important quantity in column design. However, the onset of buckling is not the end of the story for all columns—for compliant columns, it is just the beginning! Compliant columns may exhibit "postbuckling" strength and continue to carry loads after buckling. Consider the column in Figure 5.12 (this is the same as the classic nonlinear *elastica* problem—see Timoshenko and Gere, 1961).

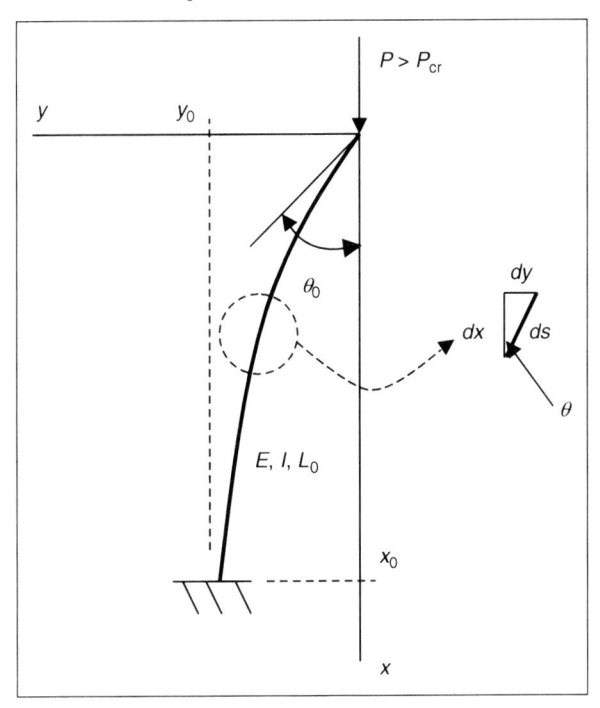

Figure 5.12. Compliant column definition sketch.

The moment–curvature relation at any point along the column is

$$EI \frac{d\theta}{ds} = -Py \tag{5.8}$$

where $d\theta/ds$ is the exact curvature (see Problems at the end of this chapter), ds is an infinitesimal measure of length, and θ measures the rotation of the column (from vertical). Recognizing that $dy/ds = \sin\theta$ and differentiating equation (5.8) with respect to ds, we get

$$\frac{d^2\theta}{ds^2} - \frac{P}{EI}\frac{dy}{ds} = -\frac{P}{EI}\sin\theta \tag{5.9}$$

Chapter 5

To solve equation (5.9), we multiply both sides by $d\theta$ and integrate

$$\frac{d^2\theta}{ds^2}d\theta = \frac{d^2\theta}{ds^2}\frac{d\theta}{ds}ds = -c^2\sin\theta\, d\theta$$

$$\int \frac{d^2\theta}{ds^2}\frac{d\theta}{ds}ds = \frac{1}{2}\int \frac{d}{ds}\left(\frac{d\theta}{ds}\right)^2 ds = -\int c^2\sin\theta\, d\theta \tag{5.10}$$

where

$$c^2 = \frac{P}{EI} \tag{5.11}$$

Upon integration, equation (5.10) gives

$$\frac{1}{2}\left(\frac{d\theta}{ds}\right)^2 = c^2\cos\theta + A \tag{5.12}$$

where A is a constant of integration to be found from boundary conditions: at $y = 0$, $d\theta/ds = 0$ because $M = 0$ and also $\theta = \theta_0$. Then evaluating equation (5.12) at $y = 0$ gives

$$y = 0:\ 0 = c^2\cos\theta_0 + A \Rightarrow A = -c^2\cos\theta_0 \tag{5.13}$$

Substituting equation (5.13) into equation (5.12), we get

$$\left(\frac{d\theta}{ds}\right)^2 = 2c^2(\cos\theta - \cos\theta_0)$$

or

$$\frac{d\theta}{ds} = \pm\sqrt{2}c\sqrt{(\cos\theta - \cos\theta_0)} = f(\theta) \tag{5.14}$$

If we assume that the curvature shown in Figure 5.12 is negative (the "sign" of the curvature here is arbitrary), then the total deformed length of the column is positive and is found by solving for ds in equation (5.14) and integrating

$$L = \int ds = -\int_{\theta_0}^{0}\left(\frac{1}{f(\theta)}\right)d\theta = \int_0^{\theta_0}\frac{d\theta}{\sqrt{2}c\sqrt{\cos\theta - \cos\theta_0}}$$

$$= \frac{1}{2c}\int_0^{\theta_0}\frac{d\theta}{\sqrt{\sin^2(\theta_0/2) - \sin^2(\theta/2)}} \tag{5.15}$$

It can be shown (see Problems at the end of this chapter) that by using a new variable (say ϕ) and letting $\xi = \sin\theta_0/2$,

$$\sin\frac{\theta}{2} = \sin\frac{\theta_0}{2}\sin\phi = \xi\sin\phi \tag{5.16}$$

Then equation (5.15) becomes (noting that as θ goes from 0 to θ_0, ϕ goes from 0 to $\pi/2$)

$$L = \frac{1}{c}\int_0^{\pi/2} \frac{d\phi}{\sqrt{1-\xi^2 \sin\phi}} = \frac{1}{c}K(\xi)$$

Substituting now in equation (5.11) gives

$$P = K(\xi)^2 \frac{EI}{L^2} \qquad (5.17)$$

where $K(\xi) = cL$ is the "complete elliptic integral of the first kind." $K(\xi)$ depends only on $\xi = \sin \theta_0/2$ and can be found in table form (see Appendix 4). (For small θ_0, equation (5.17) reduces to equation (5.7). See Problems at the end of this chapter.) As the magnitude of P increases, so also do θ_0 and $K(\xi)$.

Example 5.1

Consider a compliant column loaded past the critical buckling load such that if $\theta_0 = 60°$, $\xi = \sin 30° = 1/2$, $K(1/2) = 1.6867$, and

$$L = 1.687\sqrt{\frac{EI}{P}}$$

or

$$P = 2.842\frac{EI}{L^2}$$

Recall that the linear, small deformation result (equation (5.7)) was

$$P_{cr} = \frac{\pi^2 EI}{4L^2}$$

Then

$$\frac{P}{P_{cr}} = \frac{4(2.842)}{\pi^2} = 1.15$$

That is, a load 15% greater than the critical Euler buckling load will result in a postbuckling deformation where $\theta_0 = 60°$ (the slope or tangent at $y = 0$ makes a 60° angle with the vertical). This is of course independent of considerations of column strength or local buckling. (The compliant column considered here could be a plant stem; see Niklas, 1992.)

Example 5.2

It can be shown that $y_0 = 2\xi/c$. Now if $\theta_0 = 60°$, $\xi = 1/2$, $K(1/2) = 1.687 = cL$, and

$$y_0 = \frac{2(1/2)}{(K(1/2))/L} = 0.593L$$

If θ_0 is small, $\xi = \sin \theta_0/2$ is also small, and $\xi^2 \sin^2 \phi \ll 1$, so equation (5.17) becomes

$$L = \frac{1}{c}\int_0^{\pi/2} d\phi = \frac{\pi}{2c} = \frac{\pi}{2}\sqrt{\frac{EI}{P}}$$

or

$$P = P_{cr} = \frac{\pi^2 EI}{4L^2}$$

which is the Euler critical buckling load equation (5.7).

5.3.2 Beam Bending

The bending of a beam is a very close cousin to the buckling of a column studied earlier. The essential difference is that the beam carries transverse loads. Just as for the column, the scale of deformation is the distinguishing feature between the stiff and compliant beam response models. This explicitly shows up in whether we can use an approximation to the curvature in our model.

Consider the tip-loaded prismatic homogeneous elastic cantilever beam of length L_0, area moment of inertia I, and elastic modulus E shown in Figure 5.13 (note that P is not a "follower" force but remains in the z direction):
From the FBD (in the deformed configuration), we know that $M = Px$. Just as with the column, the governing moment–curvature relation is

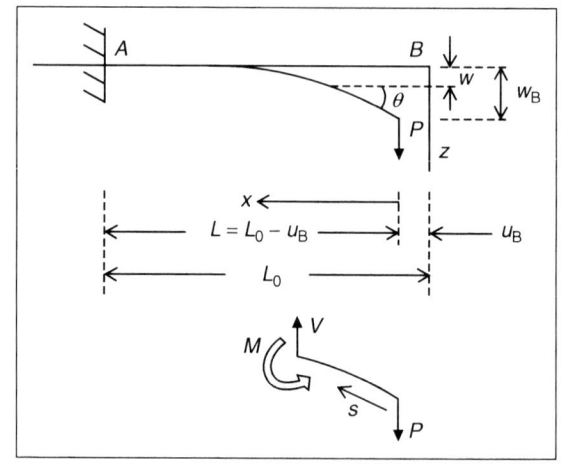

Figure 5.13. Definition sketch of a tip-loaded prismatic cantilever beam shown in the deformed configuration. The beam is represented by its neutral axis. At the lower right is a free-body diagram (FBD).

$$\text{Curvature} = -\frac{M}{EI}$$

or most generally

$$\frac{w''}{[1+(w')^2]^{3/2}} = -\frac{Px}{EI} \tag{5.18}$$

where the $'$ indicates d/dx and $''$ indicates d^2/dx^2. Integrating equation (5.18) once gives (see Problems at the end of this chapter)

$$\frac{w'}{[1+(w')^2]^{1/2}} = -\frac{Px^2}{2EI} + C_1 \tag{5.19}$$

The integration constant C_1 may be evaluated by applying the BC that $w'(L) = w'(L_0 - u_B) = 0$ (see Problems at the end of this chapter). Then equation (5.19) becomes

$$\frac{w'}{[1 + (w')^2]^{1/2}} = \frac{P}{2EI}[x^2 + (L_0 - u_B)^2] = G(x)$$

Solving (see Problems at the end of this chapter):

$$w'(x) = \frac{G(x)}{[1 - (G(x))^2]^{1/2}} \tag{5.20}$$

Integrating equation (5.20) will give the (large) deflection $w(x)$ for any $0 \le x \le L$ satisfying the BC that $w(L) = w(L_0 - u_B) = 0$. However, equation (5.20) is a function of the heretofore unknown u_B, which may be found from integrating

$$ds = \sqrt{dx^2 + dz^2} = dx\sqrt{1 + (z')^2}$$

where $z = w_B$. Integrating along the full length of the chain gives

$$\int ds = L_0 = \int_0^L \sqrt{1 + (w'_B)^2}\, dx \tag{5.21}$$

In an iterative fashion, assume u_B in equation (5.20), then carry out the integration in equation (5.21), repeat until the correct initial length is determined.

Note that equation (5.18) could also be written as

$$\frac{d\theta}{ds} = -\frac{Px}{EI} \tag{5.22}$$

Given that $dx = ds \cos\theta$ and $x = s\cos\theta$,

$$\frac{d}{ds}(\text{equation (5.22)}) = \frac{d^2\theta}{ds^2} = -\frac{P\cos\theta}{EI}$$

$$\int \frac{d^2\theta}{ds^2}\, d\theta\, \frac{ds}{ds} = -\int \frac{P\cos\theta}{EI}\, d\theta$$

$$\int \frac{d^2\theta\, d\theta}{ds^2\, ds}\, ds = -\frac{P}{EI}\int \cos\theta\, d\theta$$

$$\frac{1}{2}\int \frac{d}{ds}\left(\frac{d\theta}{ds}\right)^2 ds = -\frac{P}{EI}\int \cos\theta\, d\theta$$

$$\frac{1}{2}\left(\frac{d\theta}{ds}\right)^2 = -\frac{P}{EI}\sin\theta + C_2$$

At $x = 0$, $d\theta/ds = 0$ (since $M = 0$), and letting $\theta = \theta_0$ at $x = 0$, then

$$C_2 = \frac{P}{EI}\sin\theta_0$$

Chapter 5

Finally

$$\left(\frac{d\theta}{ds}\right)^2 = 2\frac{P}{EI}(\sin\theta_0 - \sin\theta)$$

$$\frac{d\theta}{ds} = \sqrt{2\frac{P}{EI}}\sqrt{(\sin\theta_0 - \sin\theta)}$$

$$dz = ds\sin\theta = \frac{\sin\theta\, d\theta}{(\sqrt{2P/EI})(\sqrt{\sin\theta_0 - \sin\theta})} \qquad (5.23)$$

$$z_B = w_B = \sqrt{\frac{EI}{2P}}\int_0^{\theta_0}\frac{\sin\theta\, d\theta}{\sqrt{\sin\theta_0 - \sin\theta}}$$

Now a solution procedure can be given as below:

1. Make a guess for u_B in equation (5.21), say 10% of the linear w_B and solve equation (5.21) numerically—or repeat earlier iterative process.
2. Solve for the correct w_B in equation (5.20).
3. Solve equation (5.23) numerically.

Example 5.3

Consider a homogeneous elastic, prismatic cantilever beam of initial length $L_0 = 1,000$ in. (25.4 m), $EI = 180,000$ kip in.^{2}, and a tip load P ranging from 0 to 1 kip (0–4.45 kN). Determine the corresponding tip deflection and slope for both the stiff and compliant beams.

Variables defined in MathCAD:

$$L_0 = 1,000 \text{ in.}, \quad EI = 180,000 \times 10^3 \text{ lb in.}^2, \quad P = 0.2 \times 10^3 \text{ lb}$$

Compute the linear tip displacement and guess u_B:

$$w_B = \frac{PL_0^3}{3EI}$$

$$w_B = 370.37 \text{ in.}$$

$$u_B = 0.1\, w_B$$

Find the actual u_B using equation (5.21).

Given that

$$(L_0 - u_B)\left[\int_0^1 \frac{1}{[1 - [((P(L_0 - u_B)^2)/2EI) \times (\xi^2 + 1)]^2]^{1/2}} d\xi\right] = 1{,}000$$

we find $u_B = 164.297$ in.

Find the tip angle θ_0 in degrees using equation (5.20):

$$w = \frac{(P/2EI)[x^2 + (L_0 - u_B)^2]}{[1 - [(P/2EI)[x^2 + (L_0 - u_B)^2]]^2]^{1/2}}$$

Slope $= \dfrac{180}{\pi} a\tan(w)$

Slope $= 22.83$

Table E.5.1 shows a few results of the solution:

Table E.5.1.

Load P (kip)	Linear	Nonlinear		
	w_B (in.)	w_B (in.)	u_B (in.)	θ_0 (degrees)
0.2	370.4	228.5	164.3	22.83
0.4	555.6	245.3	258.4	27.28
0.6	740.7	238.1	335.8	29.35
0.8	925.9	220.3	400.1	29.99

5.3.3 Hanging Cable

Figure 5.14 shows a uniform chain or cable of initial length L_0 hanging from two supports under its own weight $\mu g L_0$, where μ is the mass/length of the cable and g is the gravitational acceleration. The two supports are at the same level and span a distance L.

The chain is perfectly compliant in flexure but inextensible, i.e., if the chain has an unloaded length L_0, then the length under load is also L_0. How can this be—this combination of compliance but inextensibility? Consider the link chain shown in Figure 5.15. Imagine the chain only lightly loaded so that the individual link deformation is negligible, yet the link connections allow great freedom of motion.

Chapter 5

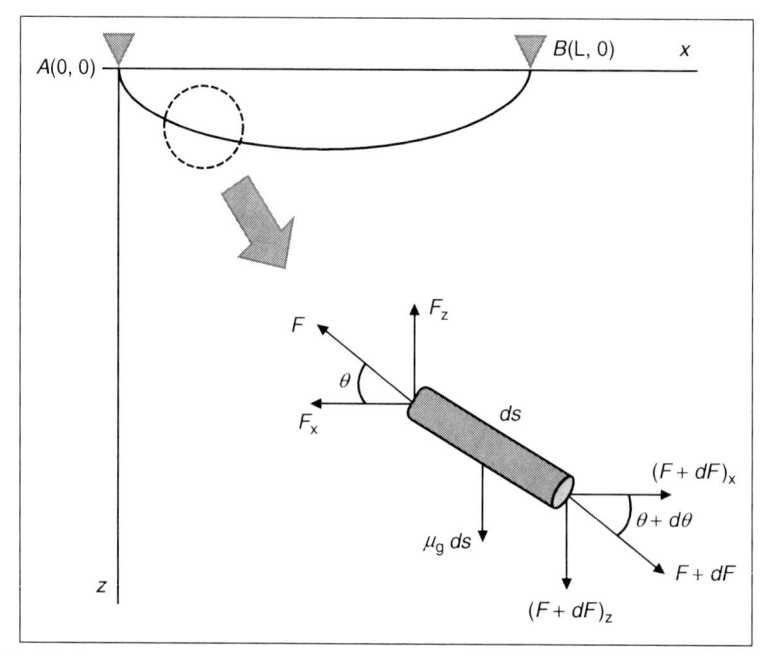

Figure 5.14. Compliant chain definition sketch. In the figure, ds is an increment of arc length along the chain, and

$$F + dF = F + \left(\frac{dF}{ds}\right)ds$$

$$F_x = F\cos\theta = F\frac{dx}{ds}, \quad (F + dF)_x = F_x + \left(\frac{dF_x}{ds}\right)ds$$

$$F_z = F\cos\theta = F\frac{d_z}{ds}, \quad (F + dF)_z = F_z + \left(\frac{dF_z}{ds}\right)ds$$

Figure 5.15. Close up view of a link chain.

Compliant Structures

A result of the cable's perfect flexural compliance is that the cable develops internal forces that are only tangential to the cable (see Figure 5.14). Those tangential forces are then resolved into components along the coordinate directions. Equilibrium in the vertical direction (z direction) requires that

$$+\downarrow \sum F_z = 0: -F\frac{dz}{ds} + F\frac{dz}{ds} + \frac{d}{ds}\left(F\frac{dz}{ds}\right)ds + \mu g\, ds = 0$$

or

$$\frac{d}{ds}\left(F\frac{dz}{ds}\right) = -\mu g \qquad (5.24)$$

Likewise in the horizontal direction (x direction)

$$+\rightarrow \sum F_x = 0: -F\frac{dx}{ds} + F\frac{dx}{ds} + \frac{d}{ds}\left(F\frac{dx}{ds}\right)ds = 0$$

or

$$\frac{d}{ds}\left(F\frac{dx}{ds}\right) = 0 \Rightarrow F\frac{dx}{ds} = \text{constant} \qquad (5.25)$$

Combining equation (5.25) with equation (5.24), and using

$$\frac{d}{ds} = \frac{d}{dx}\frac{dx}{ds}$$

and

$$\frac{dz}{ds} = \frac{dz}{dx}\frac{dx}{ds}$$

gives

$$\frac{d}{dx}\left(F\frac{dz}{dx}\frac{dx}{ds}\right) = -\mu g\frac{ds}{dx}$$

or

$$F_x\frac{d^2z}{dx^2} = -\mu g\frac{ds}{dx} \qquad (5.26)$$

(The right-hand side of equation (5.26) represents the "intensity" of the load per unit "span" length. If that load is constant, we have a model of a suspension bridge. In that case, equation (5.26) results in a parabolic profile for the chain, i.e.,

$$\frac{d^2z}{dx^2} = \text{constant} \Rightarrow z = ax^2 + b$$

Recognizing that $ds^2 = dx^2 + dz^2$ and thus $(ds/dx) = \sqrt{1 + (dz/dx)^2}$, equation (5.26) becomes

$$F_x\frac{d^2z}{dx^2} + \mu g\sqrt{1 + \left(\frac{dz}{dx}\right)^2} = 0 \qquad (5.27)$$

Chapter 5

Stiff (Linear) Model

In the linear model, the central "sag" (z at $L/2$) would be small, and hence the slope of the cable would also be small. In that case, $(dz/dx)^2 \ll 1$ and equation (5.27) becomes

$$F_x \frac{d^2 z}{dx^2} + \mu g = 0$$

Integrating once

$$\frac{dz}{dx} = -\frac{\mu g}{F_x} + C_1$$

and again

$$z(x) = -\frac{\mu g}{2 F_x} x^2 + C_1 x + C_2 \tag{5.28}$$

Applying boundary conditions $z(0) = 0$ and $z(L) = 0$ gives the following stiff response (see Problems at the end of this chapter):

$$z(x) = -\frac{\mu g}{2 F_x}(x^2 + Lx) = \frac{\mu g L^2}{2 F_x}\left[\frac{x}{L} - \left(\frac{x}{L}\right)^2\right] \tag{5.29}$$

Thus, the cable has a parabolic profile when the deflection under self-weight is small.

Compliant (Nonlinear) Model

Returning now to equation (5.27), it can be shown that a solution would have the form (Irvine, 1992)

$$z(x) = \frac{F_x}{\mu g}\left[\cosh \frac{\mu g L}{2 F_x} - \cosh \frac{\mu g}{F_x}\left(\frac{L}{2} - x\right)\right] \tag{5.30}$$

The substitution of equation (5.30) into the integration of ds then results in (see Problems at the end of this chapter)

$$s = \int_0^x \sqrt{1 + \left(\frac{dz}{dx}\right)^2}\, dx = \frac{F_x}{\mu g}\left[\sinh \frac{\mu g L}{2 F_x} - \sinh \frac{\mu g}{F_x}\left(\frac{L}{2} - x\right)\right] \tag{5.31}$$

For $s = L_0$

$$\frac{\mu g L_0}{2 F_x} = \sinh \frac{\mu g L}{2 F_x} \Rightarrow \text{solve for } F_x \tag{5.32}$$

The profile described by the hyperbolic cosine in equation (5.30) is called a *catenary*. The catenary revolving around its axis forms a "minimal surface" (a *catenoid*). When $\mu g L/F_x$ is small, the sag is small and $L \approx L_0$. In this case, the cosh expanded as a power series gives the parabolic solution seen in equation (5.28).

Example 5.4
Consider a hanging chain 20 m in length from parallel supports 15 m apart. The weight per unit length equals 10 N/m. Determine the horizontal force F_x and plot the stiff and compliant profiles.

Defining the variables in MathCAD:

$$L_0 = 20 \text{ m}, \quad L = 15 \text{ m}, \quad \mu_g = 10 \text{ N/m}$$

Solving first for F_x (using the "find" solver in MathCAD):

Guess

$$F_x = 50 \text{ N}$$

Given

$$\sinh\left(\frac{\mu_g L}{2F_x}\right) - \frac{\mu_g L_0}{2F_x} = 0$$

Find $F_x = 55.504 \text{ kg m/s}^2$

$$F_x = 55.5 \text{ N}$$

Then plotting the two responses

$$z(x) = \frac{-F_x}{\mu_g}\left[\cosh\left(\frac{\mu_g L}{2F_x}\right) - \cosh\left[\frac{\mu_g}{F_x}\left(\frac{L}{2} - x\right)\right]\right], \quad z_p(x) = \frac{-\mu_g L^2}{2F_x}\left(\frac{x}{L} - \frac{x^2}{L^2}\right)$$

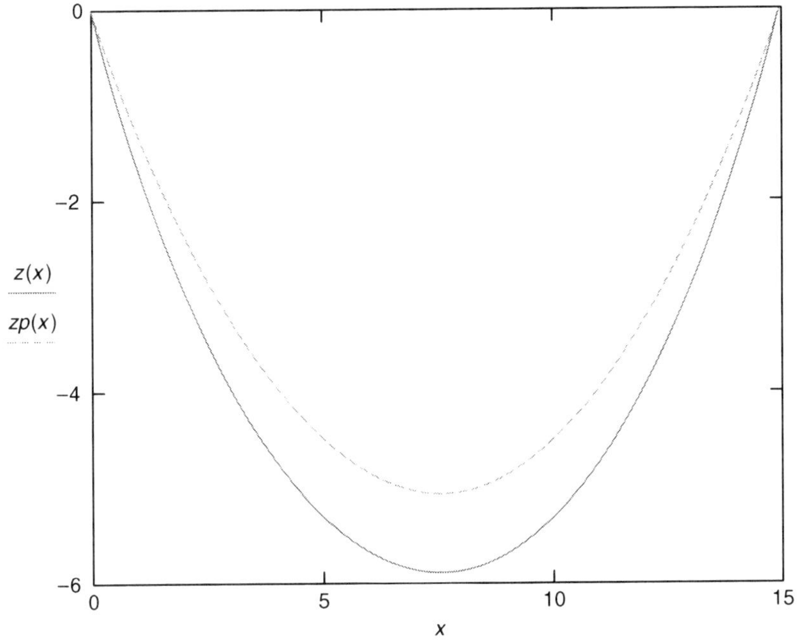

Figure E.5.1. Stiff and compliant hanging cables.

Note that the maximum sags given by the two models are
$$z(0.5\,L) = -5.887 \text{ m}, \quad z_p(0.5\,L) = -5.068 \text{ m}$$

KEY POINTS TO REMEMBER

- Structural compliance (the opposite of stiffness) is a hallmark of biological structures.
- A range of structural compliance exists, from the membrane and cable, to compliant beams, columns, and shells.
- Structural compliance results from some combination of geometry, boundary conditions, and intrinsic material properties.
- Stiff structures exhibit very small deformation under service loading and are well represented by a linear model. Compliant structures exhibit large deformation under service loads and require a nonlinear model to predict response.

REFERENCES

Irvine, M. 1992. *Cable Structures*. Dover.

Jenkins, C. H., ed. 2005. *Compliant Structures in Nature and Engineering*. WIT Press. doi: 10.2495/978-1-85312-941-4; http://library.witpress.com/pages/listpapers.asp?q_bid5408.

Lipowsky, R. 1991. "The Conformation of Membranes." *Nature* 349 (February):475–481. doi: 10.1038/349475a0; http://www.nature.com/nature/journal/v349/n6309/abs/349475a0.html.

Niklas, K. J. 1992. *Plant Biomechanics*. University of Chicago Press.

Timoshenko, S. P., and J. M. Gere. 1961. *Theory of Elastic Stability*. McGraw-Hill.

PROBLEMS

5.1. For the elastica relation $L = K(\xi)/c$,
   a. Show that for small tip angle u_0, the relation reduces to the Euler buckling formula.
   b. The elliptic integral can be represented by the power series
   $$K(\xi) = \frac{\pi}{2} \sum_{n=0}^{\infty} \left[\frac{(2n)!}{2^{2n}(n!)^2} \right]^2 \xi^{2n}$$

   Compare the power series result to the values in the table for $\theta_0 = 0°$ and $60°$, using the only first four terms of the series.

5.2. Complete the following table for the elastica problem:

$\theta_0 =>$	0	20	40	60	80	100	120	140
P/P_{cr}	1			1.152		1.518	1.884	2.541
y_0/L	0			0.593				

Plot y_0 vs. θ_0 and comment on the physical meaning.

5.3. The flower stalk of a plant has a prismatic circular cross section (Niklas, 1992). Typical values for the stalk are $d = 6$ mm and $E = 3.55 \times 10^8$ Pa, and the average weight of the flower cluster is 6.57×10^{-2} N.

   a. Compute the stalk length for a perfectly vertical stalk ($\theta_0 = 0$).
   b. The average stalk length measurement in the field is 0.50 m. Explain any difference with the value found in part (a).
   c. Using the measured stalk length of 0.50 m, what is the predicted maximum load the stalk can withstand and remain vertical? Compare this value with the flower load and discuss any difference.

5.4. For the compliant beam,

   a. Show that by using equation (3), equation (4) can be written as

   $$L_0 = \int_0^L \frac{1}{\sqrt{1 - [G(x)]^2}} dx$$

   b. Then show that a change of variables $\xi = x/(L_0 - u_B)$ leads to the first integral in the MathCAD solution file.
   c. Complete the example begun in class (complete the table) and plot the tip deflection w_B vs. load P for both the stiff and compliant solutions.

5.5. Consider a tree branch as a compliant beam

   a. Starting with a 1,000-mm-long branch ($P = 0.2$ kN, $EI = 180,000$ kN°mm^2), determine the branch length limit according to the compliant beam model. You might find it useful to fill in the following table:

L_0 (mm)	u_B (mm)	u_B/L_0 (%)	$L_0 - u_B$ (mm)	θ_0 (degrees)
1,000				
3,000				

5.6. Consider a hanging cable 20 m in length, with $\mu_g = 1.0$ N/m.

   a. For a given support span l, what is the maximum center displacement $z(x = l/2)$?
   b. What is a reasonable estimate for the maximum horizontal reaction force F_x?

c. Starting with $l = 19$ m, fill in the following table and plot $z_{parabolic}(x = l/2)$ and $z_{catenary}(x = l/2)$ vs. $L_0 - l$. Discuss when the stiff and compliant models deviate from one another.

l (m)	$L_0 - l$ (m)	$z_{parabolic}$ $(x = l/2)$ (m)	$z_{catenary}$ $(x = l/2)$ (m)	% Difference	F_x (N)
20	0	0	0	0	—
19					
18					
17					
16					
15					

5.7. The fundamental vibration frequency of a stiff cable is given by $f = c/2L_0$, where $c = (F/\mu)^{1/2}$. For the cable in Problem 5.6, determine an approximate value for f.

CHAPTER 6

Smart Structures

> Knowing a great deal is not the same as being smart.
>
> —*Carl Sagan*

6.1 INTRODUCTION

To begin, there are no engineering structures that are intrinsically "smart." We still have to define "smart," which we will do shortly, but probably the best way to start is with a definition for a *smart system*. This chapter will provide a brief introduction to biological smartness and its applications to smart engineering structures.

A smart system is one that can sense, decide, and then respond, autonomously or semiautonomously, to optimize its performance. These three actions—*sense*, *decide*, and *respond*—are the essential characteristics of an engineered smart system, structure, and so on. These actions should also convince us that engineered "smartness" requires a systems approach. For example, *responding* likely requires some electromechanical actuator, while *deciding* includes diagnosis and prognosis and usually requires some sort of electronics, be it a small digital chip or a sophisticated computer network.

Ideally, the sense/decide/respond happens without human intervention, that is, *autonomously*. But for practical reasons, depending on the degree of complexity we wish to achieve and the maturity of the technology we have to work with, we may find the need for some human intervention in the actions (hence *semiautonomously*).

As a specific example, consider a *smart structure*, which often comprises a part of, or even a great extent of, a smart system. (We might say more completely: a *smart structural system*.) Then a smart structure is a structural system that can sense, decide, and respond (semi)autonomously to optimize its load-carrying ability. Smart structures may be referred to by other names, such as adaptive structures, active structures, intelligent structures, or multifunctional structures (Srinivasan and McFarland, 2000).

Chapter 6

So-called *smart materials* are materials that "enable" smart structures or smart systems. In light of sense/decide/respond, the smartness of a material alone can only be very rudimentary. Given these caveats, at least four distinct classes of smart materials can be identified:

1. *Piezoelectrics*: The piezoelectric effect describes materials that when strained generate a voltage (actually a charge) or that deform under an applied voltage. Piezoelectrics date from the late 1800s and include the common PZT ceramic and PVDF polymer.

2. *Electroactive polymers (EAPs)*: EAPs underwent development after the Second World War, in the mid-1940s to 1950s, and include the electrostrictive polymers and the ionic polymers.

3. *Shape memory alloys/polymers (SMA/SMP)*: Shape memory metals, the best known of which are the nickel–titanium alloys (nitinol), date from the mid-1960s. More recently, the shape memory effect has been shown to exist in some polymers, although the mechanism (glassy/rubbery transition) is different than in metals (phase transformation).

4. *Rheological fluids*: Developed in the late 1940s, rheological fluids change their mechanical properties (notably yield strength) in the presence of an electric field (electrorheological fluid or ERF) or a magnetic field (magnetorheological fluid or MRF).

Keep in mind that the degree of structural complexity runs the gamut from passive and unifunctional to active, integrated, and multifunctional (Figure 6.1).

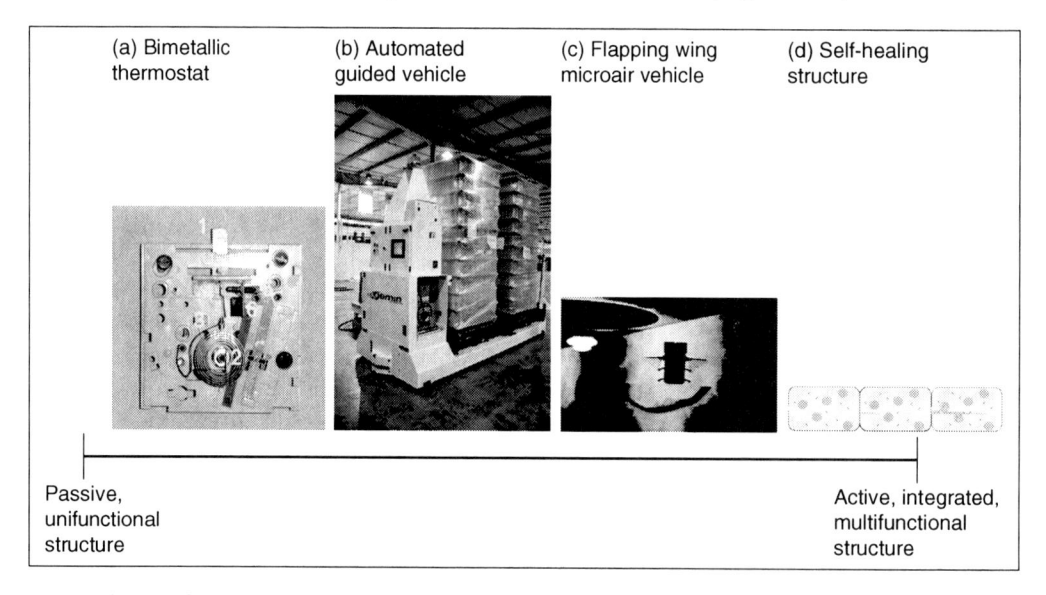

Figure 6.1. The range of smart structure complexity: (a) bimetallic thermostat (http://en.wikipedia.org/wiki/File:WPThermostat.jpg); (b) laser-guided automated guided vehicle (http://en.wikipedia.org/wiki/File:Unitload.jpg); (c) a flapping wing microair vehicle as proposed by the US Air Force, shown in a computer simulation (http://en.wikipedia.org/wiki/File:Bug_Sized_Spies,_US_air_force.jpg); (d) self-healing structure (http://en.wikipedia.org/wiki/File:Polymerization_in_situ.jpg).

As an example of a simple system, one that includes a smart structure, consider the bimetallic thermostat shown in Figure 6.2.

In Figure 6.2, the bimetallic helical strip (2) senses the ambient temperature, then "decides" to make or break the circuit (between 3 and 6), turning on or off some heating or cooling equipment. This is a rudimentary binary decision, i.e., either off or on, based on the extent of thermal deformation in the bimetal strip.

At the other extreme, we find self-healing systems and structures. Human-enacted repair of systems and structures is common and appropriate and is typically the cheapest and most effective repair method available today. However, there is great interest in systems and structures that

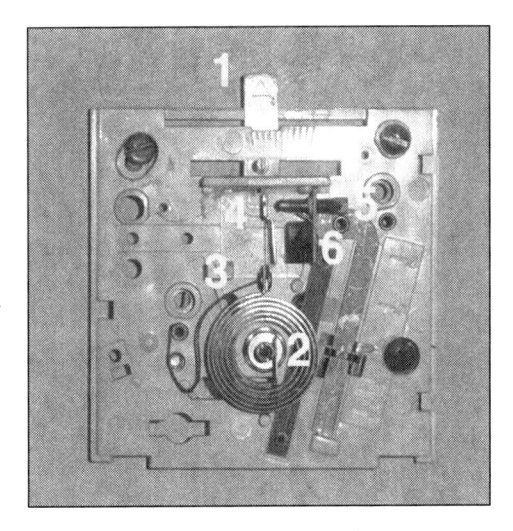

Figure 6.2. The bimetallic thermostat.
http://en.wikipedia.org/wiki/File:WPThermostat.jpg.

can heal themselves, especially where it is difficult or dangerous (and expensive!) to send in the repair person, such as in space or war zones. Some systems today do have the capacity for limited self-repair. Examples include information and power systems that can autonomously reroute around damage, and industrial robots that can retrain themselves to accommodate for conveyor variations. Self-healing structures will be discussed toward the end of this chapter.

Throughout this discussion, the connection between smart systems and biology should be obvious. Clearly, the terms smart, intelligent, adaptive, and the like, are words in the biology lexicon, as are sense, decide, and respond.

6.2 BIOLOGICAL SMARTNESS

"Smartness," at least as we shall define it, exists to one degree or another in all biological organisms. We do, however, make the distinction between smartness and intelligence, the difference being related to consciousness as we'll see below. Both smartness and intelligence can be paradigms for engineered smart structures.

The actions of sensing, deciding, and responding are the essential elements of biological smartness and intelligence. The sensing, deciding, and responding may be "intrinsic" ("hard-wired") smartness. In that case, the "deciding" is often very simply, often binary. An example is the reaction when you place your hand on a hot stove. Smartness can thus be traced to the cellular level, e.g., neurons.

Other decisions are made consciously, through *thinking*. After intrinsically removing your hand from a hot stove, what do you do next? Place your hand under cold water, blame someone, hit something, . . . ? Discussions of intelligence are very complex and much is still unknown, and we will not discuss intelligence further here.

Smartness can include fixed patterns of behavior, sometimes called *fixed action patterns* (FAPs) or more commonly *reflexes*. Reflexes are fast, involuntary responses to stimuli, which occur over simple nerve pathways called *reflex arcs*. Reflexes travel these pathways in milliseconds. There are five essential components of reflex arcs:

1. The *receptor* at the end of a sensory neuron reacts to a stimulus.
2. The *sensory neuron* conducts nerve impulses toward the central nervous system (CNS).
3. The *integration center* consists of one or more synapses in the CNS.
4. A *motor neuron* conducts a nerve impulse away from the integration center to an effector.
5. An *effector* responds to the impulses by contracting (if the effector is a muscle fiber) or secreting a product (if the effector is a gland).

Reflexes can be categorized as either autonomic or somatic. *Autonomic reflexes* are not the result of conscious control and belong to the autonomic portion of the nervous system, which includes cardiac muscle and glands. *Somatic reflexes* involve activation of skeletal muscles by the somatic portion of the nervous system.

Some reflexes involve only two neurons (*monosynaptic*), one sensory and one motor. However, most reflexes involve more than two neurons (*polysynaptic*) and include the activity of *inter-neurons* (or association neurons) in the integration center.

6.3 THE BIOLOGICAL NEURON

The brain and nervous system are a collection of several billion interconnected *neurons*. Each neuron is a cell that uses biochemical reactions to receive, process, and transmit information (sense/decide/respond!). There are *receptor* or *sensory* neurons, *motor* neurons, and *inter*-neurons.

Physical stimulation, such as light (vision), sound (audition), chemical (olfaction), or pressure (touch), is carried toward the brain by sensory neurons. Motor neurons receive impulses from other neurons and transmit this information to muscles or glands. Inter-neurons are the largest group in the nervous system. They form connections between themselves and sensory neurons before transmission of control to motor neurons.

A *nerve* provides a common pathway for the electrochemical impulses that travel along the reflex arc. A neuron's *dendritic tree* is connected to thousands of neighboring neurons (Figure 6.3). When one of those neurons fires, a positive or negative charge is received by one of the dendrites. The strengths of all the received charges are added together through the processes of spatial and temporal summation. Spatial summation occurs when several weak signals are converted into a single large one, while temporal summation converts a rapid series of weak pulses from one source into one large signal. Nerves in the CNS make synaptic contact with 1,000 to 10,000 other nerves. This allows nerve cells to be hooked together in complex patterns to perform tasks benefiting the organism.

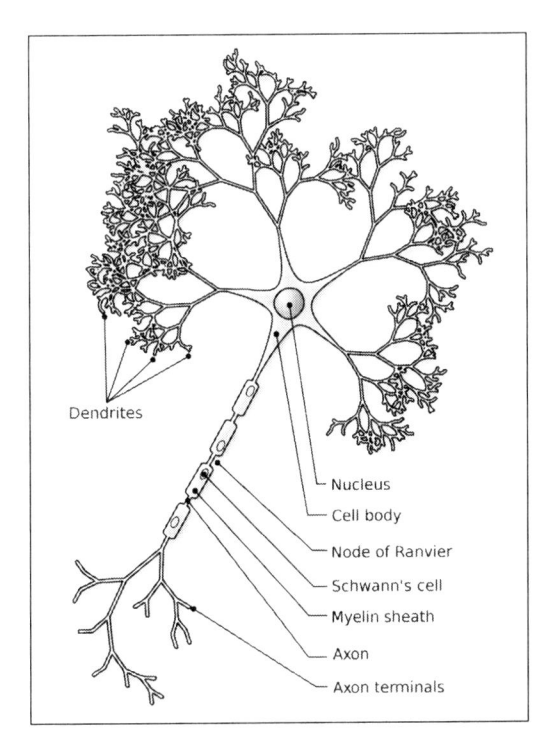

Figure 6.3. The biological neuron. http://commons.wikimedia.org/wiki/File:Neuron.jpg.

Although simple from some perspectives, biological smartness is not trivial nor is it static. For example, the vestibulo-ocular reflex, or VOR, is a reflex eye movement that stabilizes images on the retina during head movement by producing an eye movement in the direction opposite to head movement. If the VOR is poorly calibrated, then head movements result in image motion on the retina, causing blurred vision. Under such conditions, motor learning adjusts the gain of the VOR to produce more accurate eye motion. Such adjustments are needed throughout life, as neurons and muscles develop, weaken, and die—or for humans, when a new pair of eyeglasses changes the magnification of the visual field.

6.4 SELF-HEALING IN BIOLOGY AND ENGINEERING

Nature has developed a healing mechanism for organisms ranging in complexity from an amoeba (Jeon and Jeon, 1975) to lizards and starfish that are able to regenerate entire appendages. Nature's most advanced organisms, including humans and other mammals, are able to utilize large vascular networks to bring in cells from other parts of the body to help combat a wound (Clark, 1989; Morison et al., 1997).

Interestingly, there seems to be an inverse relationship between ability to heal large sections of an organism and the biological complexity of the organism. Newts have the ability to regrow appendages multiple times if an amputation of a limb occurs. This ability is possible because the specialized muscle and bone cells are able to "de-differentiate" back to a mass of stem cells that regenerate the limb.

A simple example of self-healing is the wound healing in trees, which might be called "resin infusion" (Figure 6.4). Although not truly a structural repair, the resin or sap infused into the wound does help prevent environmental degradation (moisture, insects) and at least prevents further weakening of the tree. The sap flow is more than just a simple passive, gravity-driven process but is rather governed by an intermolecular attraction (Melvin and Zimmerman, 2003).

Figure 6.4. Sap infusing into a tree's wound. http://commons.wikimedia.org/wiki/File:Tree_sap.jpg.

The stages of wound healing in higher organisms include hemostasis/inflammation, proliferation, and remodeling (Figure 6.5). The first healing responses, hemostasis (stopping blood flow in the wound area), and inflammation (swelling to a volume necessary to transport immune system components) include

1. platelet aggregation and adhesion;
2. coagulation and clotting;
3. neutrophil and monocyte/macrophage infiltration.

New tissue formation (proliferation of new cells in the wound area) proceeds quickly, followed by tissue matrix maturity (remodeling the organization of new cells in the wound area into the structure that preceded the wound event).

Given that so much of healing in nature depends on growth—the creation of new material—mimicking that miracle will always be a challenge for engineers. But we can

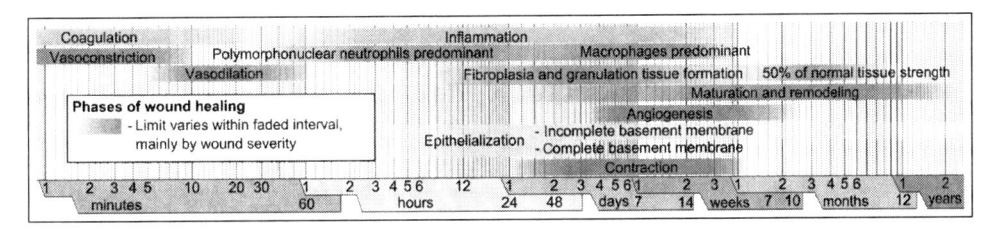

Figure 6.5. Phases of wound healing. http://en.wikipedia.org/wiki/File:Wound_healing_phases.png.

take inspiration from nature, such as the tree sap infusion shown in Figure 6.4. Pioneered by researchers at the University of Illinois, the dominant paradigm today in engineered self-healing is a "resin infusion" process wherein microcapsules of unreacted resin are embedded in a fiber–polymer matrix composite. As a crack grows, it eventually intersects and breaks a microcapsule, which releases its resin and fills the defect, at least near the crack tip (Figure 6.6). Resin curing then proceeds and a structural repair is made.

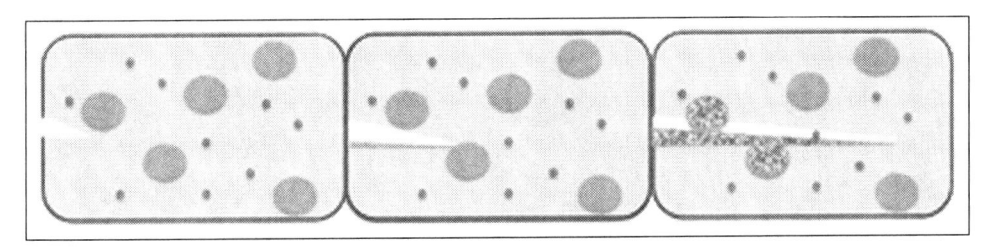

Figure 6.6. A resin infusion self-healing engineered system. http://en.wikipedia.org/wiki/File:Polymerization_in_situ.jpg.

The above described system is a passive system in that it requires no external energy to operate. Active resin infusion systems are under investigation, as are other active self-healing systems such as those using ultrasonic energy (Francischetti et al., 2011; Kellogg et al., 2010; Sarrazin et al., 2010; Zignego et al., 2011).

KEY POINTS TO REMEMBER

- The three fundamental actions of smart structural systems, in nature and engineering, are *sense/decide/respond* in order to improve performance.
- Several different classes of materials help make possible smart structures.
- In nature, we differentiate between smartness and intelligence. Smartness is hard-wired reflex, while intelligence is conscious thought.
- Smartness in biology occurs at the cellular level, in the neuron.
- The ability to self-heal is one of the hallmarks of biological systems and is being pursued by engineers for structural systems.

REFERENCES

Clark, R. A. F. 1989. "Wound Repair." *Curr. Opin. Cell Biol.* 1:1000–1008. doi: 10.1016/0955-0674(89)90072-0; http://www.sciencedirect.com/science/article/pii/0955067489900720.

Francishetti, V. S., D. L. Zignego, and C. H. Jenkins. 2011. "Ultrasonic Healing of Thermoplastic Structural Components." In *Proceedings of the 3rd International Conference on Self-Healing Materials—ICSHM 2011.* Bath, UK.

Jeon, K. W., and M. S. Jeon. 1975. "Cytoplasmic Filaments and Cellular Wound Healing in *Amoeba proteus*." *J. Cell Biol.* 67 (1):243–249. doi: 10.1083/jcb.67.1.243; http://jcb.rupress.org/content/67/1/243.

Kellogg, K. C., U. A. Korde, C. H. Jenkins, K. A. Barnes, and R. M. Winter. 2009. "On the Use of Acoustic Excitation to Accelerate Self-Healing in Polymers." In: *Proceedings of the 2nd International Conference on Self-Healing Materials—ICSHM 2009.* Chicago, IL.

Melvin, T. T., and M. H. Zimmermann. 2003. *Xylem Structure and the Ascent of Sap*, 2nd ed. Springer.

Morison, M., C. Moffatt, J. Bridel-Nixon, and S. Bale. 1997. *Nursing Management of Chronic Wounds.* Philadelphia, PA: Mosby.

Sarrazin, J. C., R. Tortop, S. Francischetti, C. H. Jenkins, U. A. Korde, and S. A. Rutherford. 2009. "Ultra-Sonic Self-Healing of Polymers." In: *Proceedings of the 2nd International Conference on Self-Healing Materials—ICSHM 2009.* Chicago, IL.

Srinivasan, A. V., and D. M. McFarland. 2000. *Smart Structures.* Cambridge University Press.

Zignego, D. L., V. S. Francischetti, and C. H. Jenkins, 2011. "Self-Healing in Thin Polymer Membranes." In: *Proceedings of the 3rd International Conference on Self-Healing Materials—ICSHM 2011.* Bath, UK.

PROBLEMS

6.1. Give three examples of "biological smartness" over a range of biological complexity. Describe the sense/decide/response activities for each.

6.2. Give two examples of intrinsic smartness in plants. Describe the sensing, the "decision," and the action.

6.3. Give five examples of intrinsic smartness in animals (other than humans). Describe the sensing, the "decision," and the action.

6.4. From your own personal experience, give five examples of intrinsic smartness in humans. Describe the sensing, the "decision," and the action.

6.5. Why is there a change in energy, between electrical and chemical, in the biological neuron? Provide analogous examples from what you know of smart structures so far.

6.6. What is intelligent life? Describe the characteristics of intelligent living systems.

6.7. Compare and contrast the bimetal thermostat with the human body's "thermostat":

   a. Describe the human body's thermostat in terms of its goals and methods of operation.

   b. Discuss the similarities and differences between the two.

6.8. Summarize applications of the following materials (one current application per category; each one appropriately referenced double-spaced paragraph plus any illustrations) in engineering problems relevant to your area of interest: (i) shape memory alloys, (ii) piezoelectric materials, (iii) electroactive polymers, and (iv) rheological fluids.

6.9. In one double-spaced paragraph with original sketches, suggest one way in which any one of the materials listed in Problem 6.8 could be applied in a problem closest to your area of interest.

CHAPTER 7

The Bio-inspired Engineering Process

Biological inspiration will lead to a new era in engineering.

—*Charles M. Vest*

Now we come to the crux of the subject: How do we *do* bio-inspired engineering? Introducing bio-inspired engineering into modern engineering design is relatively new, and there is no mature process that exists today to support that inclusion. In this chapter, we try to show how bio-inspired engineering can still be done and make important contributions to society.

7.1 FORWARD AND REVERSE DESIGN REDUX

We have seen how bio-inspired engineering (BiE) uses biological structures and functions to solve specific technical problems through inspiration from biological designs. Modern bio-inspired engineering has had a fruitful, if limited, history. BiE rests on the premise that evolutionary processes have produced biological structures and functions that are highly optimized solutions to specific problems driven by diverse environmental pressures. Most bio-inspired design relies on an approach inverse to the normal engineering design process: A biological structure–function relation is understood and then a search is conducted for an engineering problem that may be solved with that knowledge. Today, the usual forward approach (Figure 7.1), and the one most typical for design—an engineering problem in search of a biological solution paradigm—is essentially unavailable to the practicing engineer.

Chapter 7

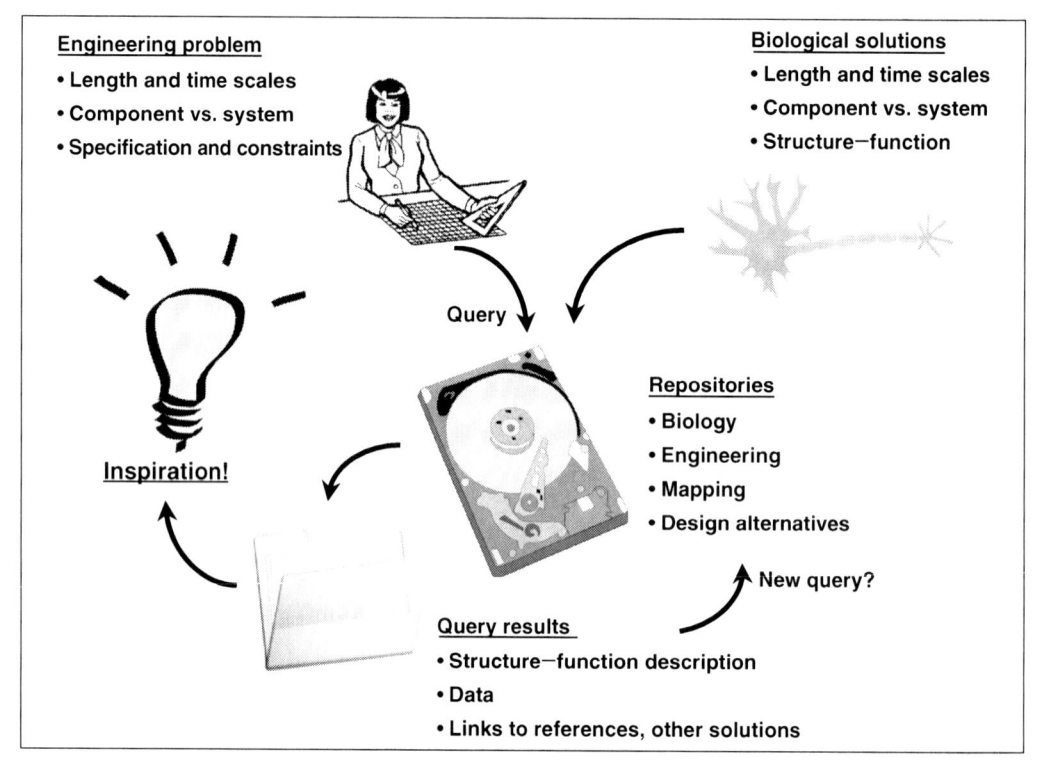

Figure 7.1. The desired forward design process in bio-inspired engineering.

The engineering design process is fairly well understood today. The usual *forward* design process begins with a set of design requirements and proceeds in an iterative fashion to a solution that (hopefully) optimally satisfies those requirements. Although minor details may vary from engineer to engineer or industry to industry, it is generally recognized that the engineering design process includes the following major steps: problem formulation, concept generation, preliminary design, detailed design, and implementation/realization. The less common *inverse* design process essentially takes a known solution, engineering or otherwise, and works backward to satisfy an open engineering problem. (More will be discussed about the inverse process later in the chapter.)

Although every step of the design process is important, once a rational set of design requirements is formulated the most critical step is concept generation (ideation). It is from the assortment of conceptual solutions that the final candidate design will be selected for detailing and ultimate realization. The quality of the final solution rests directly and indelibly with the quality of the concepts generated. There are formal processes for concept generation such as brainstorming, functional analysis, and TRIZ. TRIZ (see Section 7.2.3) is a process that provides methods and tools for generating innovative ideas and solutions. Unlike common design techniques such as brainstorming (which is founded on random idea generation), TRIZ offers a formalistic or algorithmic approach to concept generation. However, it does not directly provide access to the "catalog of biological solutions."

The natural world is a vast repository of paradigms for concept generation. Nature has had millennia to converge on solutions to very specific problems, to "get it right." We expect many biological solutions are optimized and, by definition, life sustaining. BiE focuses on learning from nature and not on learning to use nature. The fundamental limitation, however, is that the means for the engineer to access the "catalog" of natural solutions do not exist, particularly in the sense of the forward design problem. Even with well-developed design processes, like TRIZ, resources and databases are still necessary, and the engineer gets no closer to accessing the knowledge of biological solutions with any current design process.

Because of this gap, the typical BiE process is an inverse process. We considered the well-known case of Velcro in Chapter 1. In the late 1940s, Swiss engineer George de Mestral was hiking in the Alps when he observed cockleburs stuck to his clothing and to his dog's fur. Examining the seeds under a microscope, he realized that the tiny bur hooks were catching, temporarily, on fabric loops in his clothing. Although he was not initially searching for a solution to "temporary fastening," the biological solution represented by the cocklebur gave him an idea for an engineering problem he might solve. In a more recent development, Dr. Frank Fish was browsing in a Boston sculpture shop a few years ago when he noticed a whale figurine. His first thought was "This isn't right. It's got bumps on the leading edge of its flipper. It's always a straight edge." But the store manager showed him a photo of a humpback whale, and sure enough, it had bumps on its flippers. Dr. Fish speculated that the tubercles may somehow give them a hydrodynamic advantage. Fish copatented the so-called "Tubercle Technology" and in 2005 he helped found Whale Power, a company that is building energy-efficient windmills using scalloped-edge blades. The technology could eventually improve energy efficiency for any machine that uses turbines, fans, or pumps.

The two stories share several common themes: the accidental discovery of a natural design, the capacity of this design to improve on the best contemporary human technology, and the consequent design approach—from an existing natural structure to closely related engineered artifact. This illustrates quite clearly the limitation to the inverse design process, and our current inability to access the biological universe in the forward design process represents tremendous lost opportunities.

7.2 RESOURCES FOR BIO-INSPIRED ENGINEERING

The student may find the following resources useful in practicing bio-inspired engineering. They are offered here without endorsement and without attempt to be inclusive. (Text quoted is directly from the resource's website.)

7.2.1 Comprehensive Web Databases

1. AskNature (www.asknature.org). "AskNature is a free, open source project, built by the community and for the community. Our goal is to connect innovative minds with life's best ideas, and in the process, inspire technologies that create conditions conducive to life. To accomplish this, we're doing something that has never been done—organizing the world's biological literature by function. What you'll see on the site today is a starter culture of ideas—biological blueprints and strategies, bio-inspired products and design sketches, and biomimics you can talk to and collaborate with."

2. Encyclopedia of Life (www.eol.org). "Encyclopedia of Life [is] an unprecedented global effort to document all 1.8 million named species of animals, plants, and other forms of life on Earth. For the first time in the history of the planet, scientists, students, and citizens will have multi-media access to all known living species, even those that have just been discovered. The Field Museum of Natural History, Harvard University, Marine Biological Laboratory, Smithsonian Institution, and Biodiversity Heritage Library joined together to initiate the project, bringing together species and software experts from across the world. The Encyclopedia of Life will provide valuable biodiversity and conservation information to anyone, anywhere, at any time," said Dr. James Edwards, currently Executive Secretary of the Global Biodiversity Information Facility who today was officially named Executive Director of the Encyclopedia of Life. "Through collaboration, we all can increase our appreciation of the immense variety of life, the challenges to it, and ways to conserve biodiversity. The Encyclopedia of Life will ultimately make high-quality, well-organized information available on an unprecedented level. Even five years ago, we could not create such a resource, but advances in technology for searching, annotating, and visualizing information now permit us, indeed mandate us to build the Encyclopedia of Life."

3. Tree of Life (www.tolweb.org/tree). "The Tree of Life Web Project is a collection of information about biodiversity compiled collaboratively by hundreds of expert and amateur contributors. Its goal is to contain a page with pictures, text, and other information for every species and for each group of organisms, living or extinct. Connections between Tree of Life web pages follow phylogenetic branching patterns between groups of organisms, so visitors can browse the hierarchy of life and learn about phylogeny and evolution as well as the characteristics of individual groups."

7.2.2 Scholarly Journals

1. *International Journal of Design & Nature and Ecodynamics* (www.journals.witpress.com). "The International Journal of Design & Nature and Ecodynamics acts as a channel of communication for researchers from around the world working on a variety of studies involving nature and its significance to modern

scientific thought and design. These studies have demonstrated the rich diversity of the natural world."

2. *Journal of Biomimetics, Biomaterials, and Tissue Engineering* (http://www.scientific.net/JBBTE). "Journal of Biomimetics, Biomaterials, and Tissue Engineering is a peer-reviewed scientific journal dedicated to publishing research and review papers in the general areas of biomimetics, biomaterials, and tissue engineering. As such, it is a multidisciplinary journal intersecting the areas of biology, biochemistry, and materials science and engineering. While the editors will consider all manuscript submissions in these general areas, the focus of this journal is to explore the field of biomimetics and how it can be the key to scientific progress in the related areas of biomaterials and tissue engineering."

3. *Bioinspiration & Biomimetics* (www.iopscience.iop.org). "Bioinspiration & Biomimetics publishes research which applies principles abstracted from natural systems to engineering and technological design and applications. The journal will publish research involving the study and distillation of principles and functions found in biological systems that have been developed through evolution, and application of this knowledge to produce novel and exciting basic technologies and new approaches to solving scientific problems. It will provide a forum for interdisciplinary research which will act as a conduit to facilitate the two-way flow of ideas and understanding between the extensive bodies of knowledge of the different disciplines. It has two principal aims: to draw on biology to enrich engineering and to draw from engineering to enrich biology. The journal aims to include input from across all intersecting areas of both fields. In biology, this would include work in all fields from physiology to ecology, with either zoological or botanical focus. In engineering, this would include both design and practical application of biomimetic or bioinspired devices and systems."

4. *Nature Materials* (www.nature.com/nmat). "Nature Materials provides a forum for the development of a common identity among materials scientists while encouraging researchers to cross established subdisciplinary divides. To achieve this, and strengthen the cohesion of the community, the journal takes an interdisciplinary, integrated and balanced approach to all areas of materials research while fostering the exchange of ideas between scientists involved in the different disciplines. Nature Materials is an invaluable resource for all scientists, in both academia and industry, who are active in the process of discovering and developing materials and materials-related concepts. Research areas covered in the journal include: engineering and structural materials (metals, alloys, ceramics, composites); organic and soft materials (glasses, colloids, liquid crystals, polymers); and bio-inspired, biomedical and biomolecular materials."

5. *International Journal of Bio-Inspired Computation* (www.inderscience.com). "The major goal of *IJBIC* is the publication of new research results on bio-inspired computation methods and their applications. *IJBIC* provides the scientific community and industry with a vehicle whereby ideas using two or more conventional and computational intelligence based techniques can be discussed. Bio-inspired

computation is an umbrella term for different computational approaches that are based on principles or models of biological systems. This class of methods, such as evolutionary algorithms, ant colony optimization, and swarm intelligence, complements traditional techniques in the sense that the former can be applied to large-scale applications where little is known about the underlying problem and where the latter approaches encounter difficulties. Therefore, bio-inspired methods are becoming increasingly important in the face of the complexity of today's demanding applications, and accordingly they have been successfully used in various fields ranging from computer engineering and mechanical engineering to chemical engineering and molecular biology."

7.2.3 Other Resources

1. TRIZ (http://www.triz-journal.com/). "'TIPS' is the acronym for 'Theory of Inventive Problem Solving,' and 'TRIZ' is the acronym for the same phrase in Russian. TRIZ was developed by Genrich Altshuller and his colleagues in the former USSR starting in 1946, and is now being developed and practiced throughout the world. TRIZ research began with the hypothesis that there are universal principles of invention that are the basis for creative innovations that advance technology, and that if these principles could be identified and codified, they could be taught to people to make the process of invention more predictable. The research has proceeded in several stages over the last 50 years. Over 2 million patents have been examined, classified by level of inventiveness, and analyzed to look for principles of innovation."

2. Constructal Theory (www.constructal.org). "The constructal law accounts for the design of the biosphere: the global movement of mass, as the action of constructal engines (geophysical, animal, human made) that dissipate their power into brakes (interactions with the environment). The resulting designs are global: vegetation, animal locomotion, vision, cognition, hierarchy, science and technology evolution, economic activity (GNP) and the production and distribution of useful energy."

3. See also the following references:

   Ball (2001); Jenkins and Larsen, (2008); Kapsali and Dunamore (2008), Larsen et al. (2009), Menon and Lan (2006), Vincent and Mann (2002), and Vincent et al. (2005)

7.3 REVERSE DESIGN FOR BIO-INSPIRED ENGINEERING

Given that the forward design process is not practically available today for bio-inspired engineering, we rely on reverse design. We studied reverse engineering design in Chapter 2, but the application to BiE is perhaps best illustrated through example. In the following example, we use the Reverse Bio-Inspired Engineering Worksheet (Appendix 3).

Example 7.1

Reverse Bio-Inspired Engineering Worksheet
Three-Horned Rhinoceros Beetle

1. Examine the biological artifact with the intent of discerning:
   a. What does the biological artifact do?
      – *Lives day-to-day by surviving, by eating, resting, mating, and finding shelter.*

Figure E.7.1. Three-horned rhinoceros beetle. http://commons.wikimedia.org/wiki/File:Rhinoceros_beetle.jpg.

   b. How does the biological artifact work?
      – *Has six long legs equipped with hook-like structures for grip, with a large body (in comparison to other beetles) protected by thick outer wing covering a set of larger wings for flying. The beetle's head is large with three horn-like structures made for fighting and scavenging for food.*
   c. What might the biological artifact's "requirements" have been?
      (i) *Defense mechanism*
      (ii) *Offensive mechanism*
      (iii) *Form of travel*

Chapter 7

2. Relate the biological artifact's features to the artifact requirements listed in 1.c.:
   a. List the biological artifact's features (geometry, materials, mechanisms, etc.)
      (i) *Thick-shelled head and thick outer wing for body protection.*
      (ii) *Three horns made for fighting other rhino beetles and for digging.*
      (iii) *Grappling hooks of the beetle's six legs for high gripping ability, as well as spiked up parts of the legs. The beetle also has a set of wings underneath the thick set for flying, although not very well.*
   b. How do the biological artifact's features support the requirements?

Features	Requirements		
	(i) Defensive	(ii) Offensive	(iii) Travel
(i) Thick-shelled	—	Adds weight, giving the beetle a better pushing force for the beetle's sumo-style fighting	Adds a larger weight, making flying and distance travel harder
(ii) Three horns	Adds a scare factor for predators, not entirely useful for fighting off predators if they aren't sharp	—	The horns don't aid travel at all, only make it a little less convenient due to their size and weight
(iii) Legs and wings	Legs—can produce a wide stance, making the beetle harder to roll over exposing the not as protected underbelly Wings—can help get away from trouble	The legs' wide stance also helps create a strong base for fighting and trying to flip the other rhino beetle over	—

   c. Are there features that don't map to any requirements?
      Not really, the beetle is designed for necessary functions.

124

3. Form and Function
   a. How do the form (shape) and material of a feature relate to the function that the feature performs?

Features	1. Function	2. Form	3. Material	Relationship of 1.–2.–3.
(i) Thick-shelled	Protection from predators	The body is shell-like over the back and head fully covered	Exoskeleton	A strong material that is in a strong shell shape for maximum protection
(ii) Three horns	Fighting other beetles for dominance and digging for food and shelter	Two curved into each other in an x–y plane and another curved up in a y–z plane	Exoskeleton	A strong material with long reaching range and the horns angled differently for optimal grabbing
(iii) Legs and wings	For travel and movement	Six long legs and two sets of wings	Exoskeleton and thin membrane wing	Longer legs for reaching and hooking on to something to be able to travel through more challenging terrain

4. Product Improvement
   a. Suggest a new product or process based upon what you've learned in 1–3 above.
   - *The hooked feet structure is very similar to a grappling hook.*
   - *The three horns promote a grabbing arm that would be able to grab cylindrical structures of multiple sizes (Figure E.7.2).*
   - *The joints that attach the wing to the body are spring loaded—the joint unlocks and the wings pop out, opposed to forcing them out and in. This produces a much shorter time required for the beetle to be able to be ready to fly, using the speed when needed to get away from something quickly.*

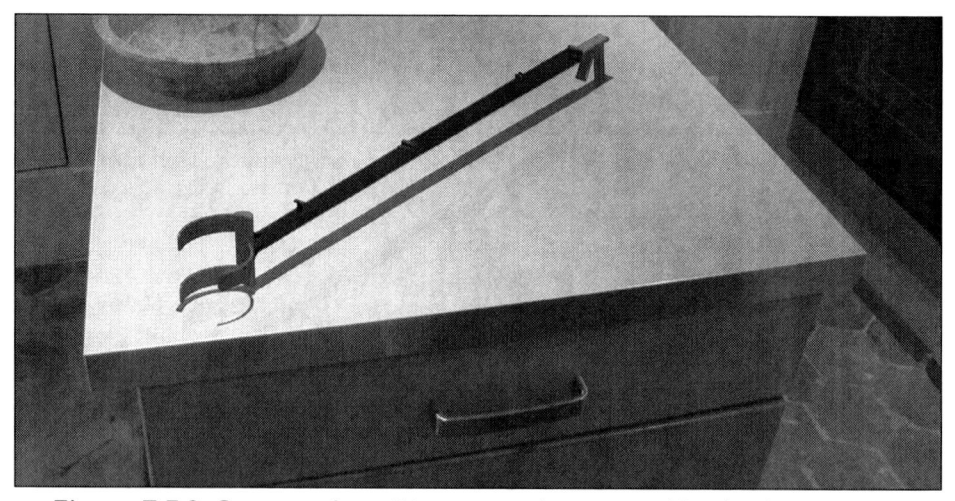

Figure E.7.2. Concept of possible new product inspired by the three-horned rhinoceros beetle.

KEY POINTS TO REMEMBER

- Even though a mature formal process does not currently exist, bio-inspired engineering can still be done.
- Many resources are available to aid in the process, but the user should apply due caution.
- A reverse engineering approach is one viable option for doing bio-inspired engineering.

REFERENCES

Ball, P. 2001. "Life's Lessons in Design." *Nature* 409:413–415.

Jenkins, C. H., and J. J. Larsen. 2008. "Deployment Schemes for 2-D Space Apertures and Mapping for Bio-Inspired Design." In *Proceedings of the 6th International Conference on Computation of Shell and Spatial Structures IASS-IACM 2008*. Ithaca, NY.

Kapsali, P., and P. Dunamore. 2008. "Biomimetic Principles in Clothing Technology." In *Biologically Inspired Textiles*, edited by A. Abbott, and M. Ellison, *Woodhead Publishing Series in Textiles No. 77*. doi: 10.1533/9781845695088; http://www.woodheadpublishingonline.com/content/xlrp0520454h1731/.

Larsen, J. J., C. H. Jenkins, and K. Woo. 2009. "A Bio-Inspired Lightweight MRF-Foam Actuator." In *10th Gossamer Spacecraft Forum, 50th AIAA/ASME/ASCE/AHS/ASC Structures, Structural Dynamics, and Materials Conference*. Palm Springs, CA.

Menon, C., and N. Lan. 2006. "Methodical Approach to Implement Biomimetic Paradigms in the Design of Novel Engineering and Space Systems." In *International Astronautical Congress, IAC-06-D3.P.03.*

Vincent, J. F. V., and D. L. Mann. 2002. "Systematic Technology Transfer from Biology to Engineering." *Phil. Trans. R. Soc. A* 360:159–173. doi: 10.1098/rsta.2001.0923; http://rsta.royalsocietypublishing.org/content/360/1791/159.

Vincent, J. F. V., O. Bogatyreva, A.-K. Pahl, N. Bogatyrev, and A. Bowyer. 2005. "Putting Biology into TRIZ: A Database of Biological Effects." *Creativ. Innovat. Manage.* 14 (1): 66–72. doi: 10.1111/j.1476-8691.2005.00326.x; http://onlinelibrary.wiley.com/doi/10.1111/j.1476-8691.2005.00326.x/abstract;jsessionid=09425BAA98FA3A4C75F16B4E900A6220.d03t02.

PROBLEMS

This is a design project used in ME 455 Bio-Inspired Engineering at Montana State University. This can be modified to suit the needs of the instructor and the student.

Chapter 7

DESIGN PROJECT

The most common procedure for bio-inspired engineering is the inverse approach (aka reverse engineering). In this project, you will work as part of a three- or four-person team to find engineering inspiration from a biological artifact.

1. Group sign-up on the sheet provided.

- *Bombus* (bumble bee)
- *Prosopocoilus giraffe* (Giraffe Stag-Beetle)
- *Chalcosoma moellenkampi* (three-horned rhinoceros beetle)
- *Cyrtotrachelus buqueti* (giant bamboo weevil)
- Sphingidae (sphinx moth)

2. Get to know your artifact! Handle it carefully. Find out as much as you can about it.
3. Reverse engineer. Refer to the "Reverse Bio-Inspired Engineering Worksheet" (Appendix 3).
4. Research all known bio-inspired applications associated directly or indirectly with your artifact to date.
5. Find an engineering problem that could be solved on the basis of inspiration from your artifact and develop the concept.

You will present your results in the form of a brief technical report and a presentation. The report must include the following sections:

- Cover page (see template below)
- Introduction (purpose of the study; what is to be found in the report)
- Biology of the artifact
- Reverse engineering study
- Relevant bio-inspired applications in the literature
- A new bio-inspired engineering solution
- Summary and conclusion

The report is to be typed double-spaced or it will be returned ungraded. Reference any materials not of your own creation, whether from books, articles, the Internet, etc. Prepare a 20-min presentation to be given to the class.

PROJECT REPORT COVER PAGE

TO: Dr. Jenkins
FROM: ME 455 Design Project Group _____
DATE:
RE: Bio-Inspired Design Project

[Project Title]
Executive Summary

Attached please find ....

The report provides the background and ......

In summary, our findings are ....

Each signature below verifies that the named person has contributed fully to the completion of this report, including proofreading of the final document.

_____   _____
Print Name   Signature and Date

_____   _____
Print Name   Signature and Date

_____   _____
Print Name   Signature and Date

_____   _____
Print Name   Signature and Date

APPENDIX I

Biological Classification

The domain of biology is vast, and biologists use classification systems to provide some order to the great diversity of the natural world. The systems used today are modern versions of early developments in the eighteenth century and are a form of *taxonomy*. The taxonomy starts broadly and proceeds to finer and finer resolution. The most common example is shown in Figure A.1.

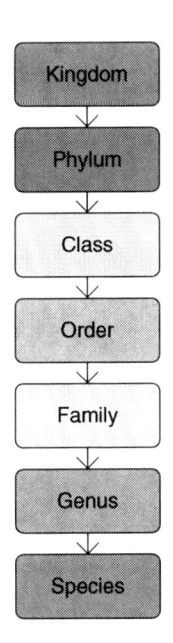

Figure A.1. The seven main ranks in biological classification.
http://commons.wikimedia.org/wiki/File:Scientific_classification.png.

Appendix 1

Biologists generally classify living organisms into five large groups or kingdoms: Monera (simple microorganisms), Protista (complex microorganisms), Fungi, Plantae, and Animalia. As an example, as humans our classification is as follows:

Kingdom = **Animalia**
Phylum = **Chordata**
Class = **Mammalia**
Order = **Primates**
Family = **Homindae**
Genus = **Homo**
Species = **Sapiens**

The "scientific name" of an organism consists of the genus and species (italicized). For example, the scientific name for humans is *Homo sapiens*.

APPENDIX 2

A Simple Inverse Mechanics Problem

Consider a single degree of freedom (SDOF) spring–mass–damper system shown in Figure A.2.1.

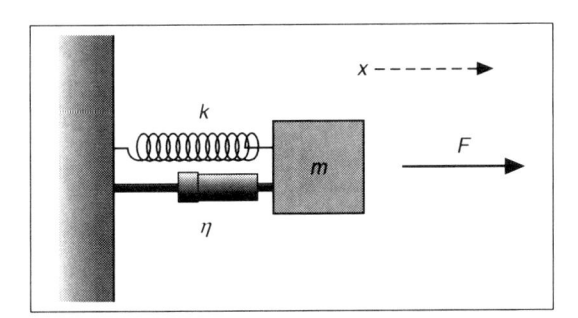

Figure A.2.1. An SDOF spring–mass–damper system. Here, k = spring stiffness, m = mass, and η = damping coefficient. This system is also called a *harmonic oscillator*.

Let the forcing on the system be simple harmonic, i.e., $F = F(t) = F_0 \sin \omega t$, where t is the time, ω is the forcing (circular) frequency (rad/s), and F_0 is the force magnitude. Then the steady-state system response is given by

$$x(t) = X(\omega) \sin(\omega t + \varphi)$$

where φ is the phase angle (lag in response due to damping) and $X(\omega)$ is the amplitude given by

$$X(\omega) = \frac{F_0/k}{\sqrt{(1 - r^2)^2 + (2\zeta r)^2}}$$

Appendix 2

and $r = \omega/\omega_n$, $\omega_n^2 = k/m$, and $\zeta = 2m\omega_n$. Figure A.2.2 shows a nondimensional $X(\omega)$ as function of r.

Case 1 is the forward problem: For a known system, what's the response? "Knowing the system" in this case means k, m, and the forcing are known, hence r is known. For a given value of r, Figure A.2.2 shows that a unique response can be determined.

Case 2 is the inverse problem: For a given response, what's the system? Figure A.2.2 shows that there are two possible "systems" (values of r), and hence there is not a unique answer. In fact, there are an infinity of answers since there are an infinity of choices for any $r = \omega/\sqrt{k/m}$

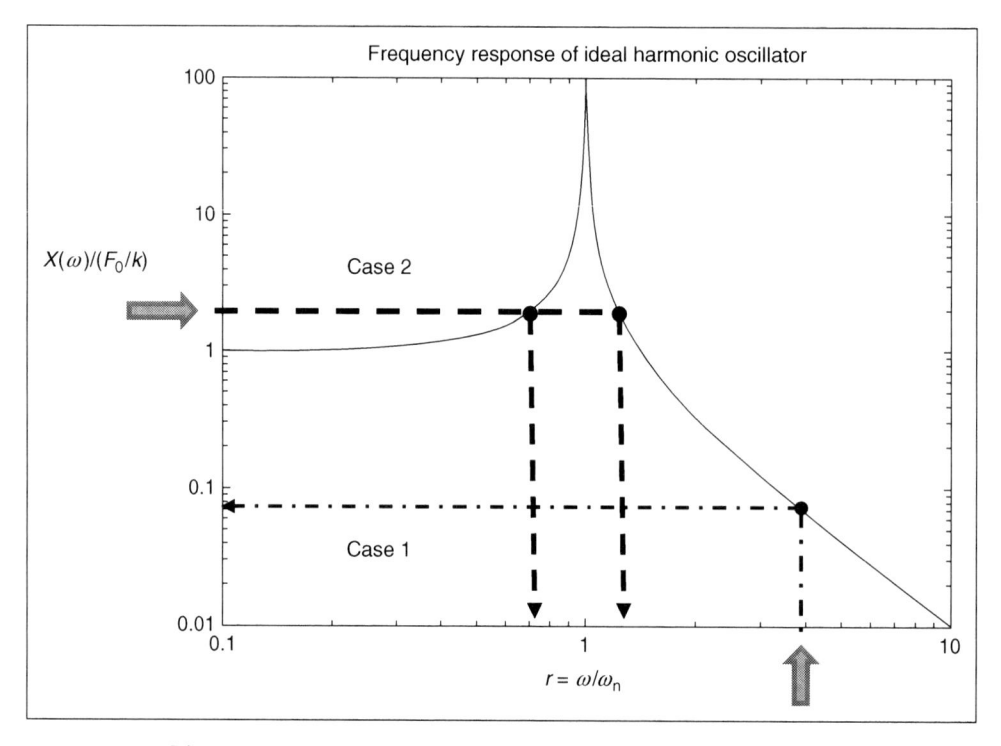

Figure A.2.2. Nondimensional $X(\omega)$ as function of $r = \omega/\omega_n$.

APPENDIX 3

Reverse Engineering Worksheets

Name _____

I. REVERSE ENGINEERING WORKSHEET

1. Examine the product with the intent of discerning:
   a. What does the product do?

   b. How does the product work?

   c. What might the product requirements have been?
      (i)

      (ii)

      (iii)

Appendix 3

2. Relate the product features to the product requirements listed in 1.c.:
   a. List the product features (geometry, materials, mechanisms, etc.).
      (i)

      (ii)

      (iii)

   b. How do product features support the requirements? (Expand the table as required.)

Features	Requirements		
	(i)	(ii)	(iii)
(i)			
(ii)			
(iii)			

   c. Are there features that don't map to any requirements?

3. Form and Function
    a. How do the form (shape) and material of a feature relate to the function that the feature performs?

Features	1. Function	2. Form	3. Material	Relationship of 1.–2.–3.
(i)				
(ii)				
(iii)				

4. Product Improvement
    a. How might the product be improved?

    b. Any inspiration for a new product?

Appendix 3

Name _____

II. REVERSE BIO-ENGINEERING WORKSHEET

1. Examine the biological artifact with the intent of discerning:
   a. What does the biological artifact do?

   b. How does the biological artifact work?

   c. What might the biological artifact's "requirements" have been?
      (i)

      (ii)

      (iii)

2. Relate the biological artifact's features to the artifact requirements listed in 1.c.:
   a. List the biological artifact's features (geometry, materials, mechanisms, etc.)
      (i)

      (ii)

      (iii)

b. How do the biological artifact's features support the requirements?

Features	Requirements		
	(i) Defensive	(ii) Offensive	(iii) Travel

c. Are there features that don't map to any requirements?

3. Form and Function
   a. How do the form (shape) and material of a feature relate to the function that the feature performs?

Features	1. Function	2. Form	3. Material	Relationship of 1.–2.–3.

4. Engineering Inspiration
   a. Suggest a new product or process based upon what you've learned in 1–3 above.

139

APPENDIX 4

Tip Angle θ_0 and the Elliptic Integral of the First Kind $K(\xi)$

θ_0 (degrees)	$K(\xi)$	θ_0 (degrees)	$K(\xi)$
0	1.57079	46	1.63651
2	1.57091	48	1.64260
4	1.57127	50	1.64899
6	1.57187	52	1.65569
8	1.57271	54	1.66271
10	1.57379	56	1.67005
12	1.57511	58	1.67773
14	1.57667	60	1.68575
16	1.57848	62	1.69411
18	1.58054	64	1.70283
20	1.58284	66	1.71192
22	1.58539	68	1.72139
24	1.58819	70	1.73124
26	1.59125	72	1.74149

(Continued)

Appendix 4

θ_0 (degrees)	$K(\xi)$	θ_0 (degrees)	$K(\xi)$
28	1.59456	74	1.75216
30	1.59814	76	1.76325
32	1.60197	78	1.77478
34	1.60608	80	1.78676
36	1.61045	82	1.79992
38	1.61510	84	1.81915
40	1.62002	86	1.82560
42	1.62523	88	1.83856
44	1.63072	90	1.85407

APPENDIX 5

Comparative Properties of Natural Materials

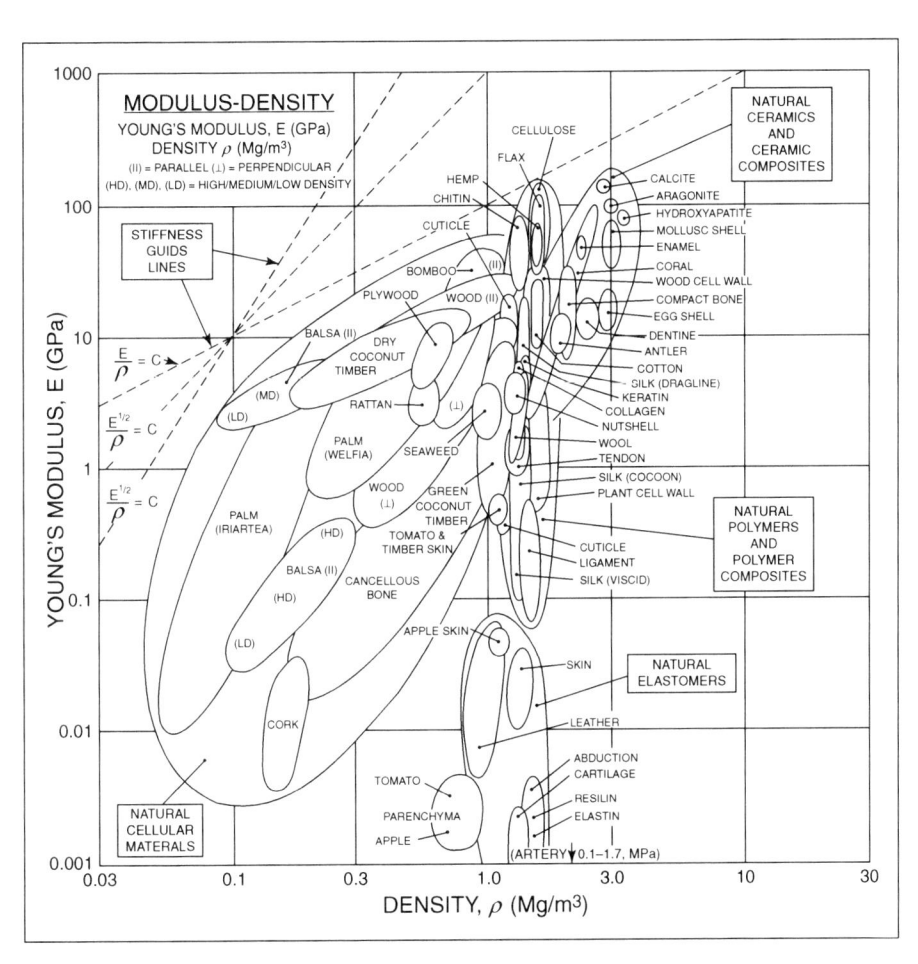

Figure A.5.1. Modulus vs. density. Reproduced with permission from Ashby, M. F. 2005. *Materials Selection in Mechanical Design*, 3rd ed., 395–397. Elsevier.

Appendix 5

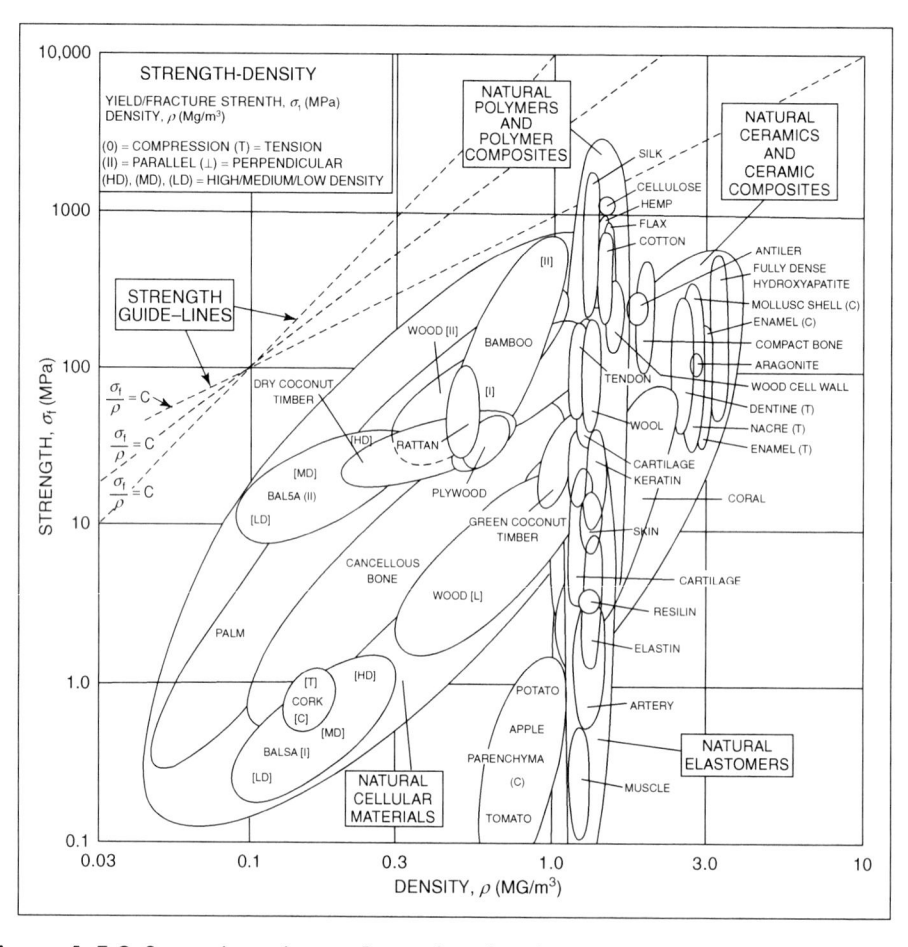

Figure A.5.2. Strength vs. density. Reproduced with permission from Ashby, M. F. 2005. *Materials Selection in Mechanical Design*, 3rd ed., 395–397. Elsevier.

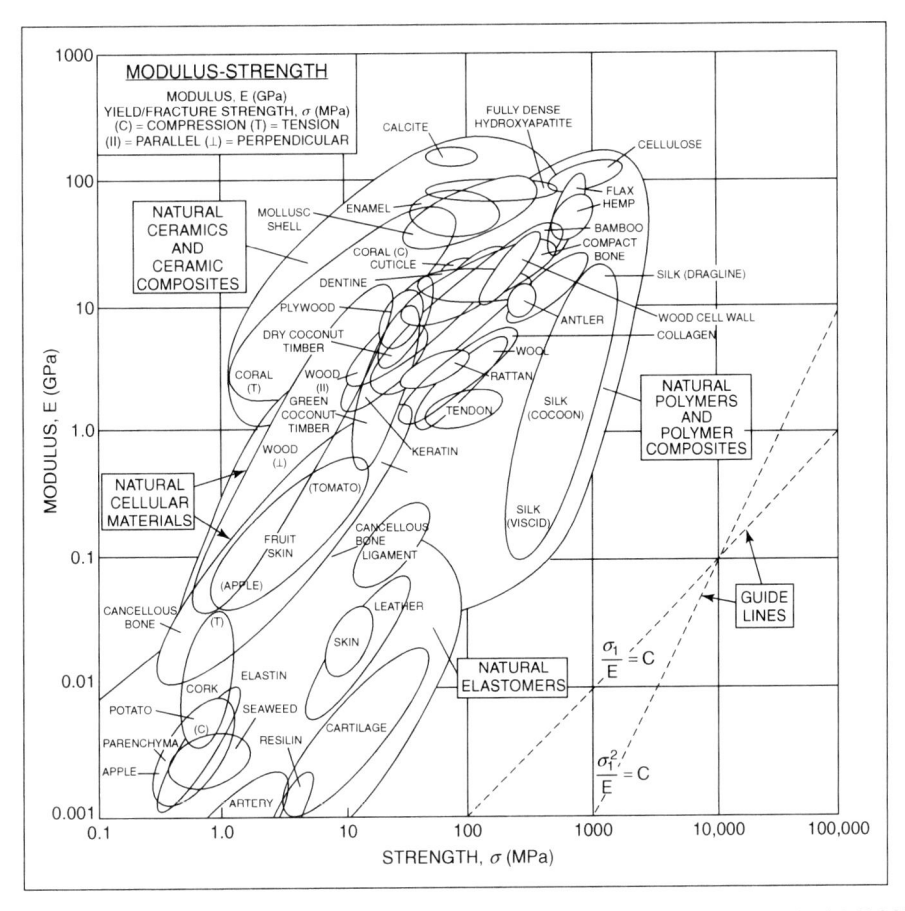

Figure A5.3. Modulus vs. strength. Reproduced with permission from Ashby, M. F. 2005. *Materials Selection in Mechanical Design*, 3rd ed., 395–397. Elsevier.

Figure Attributions

[URL citation in caption—Permission note numbers are in parens (*x*) in the chapter lists; the language, in note number order, follows.]

Chapter 1
Figure 1.1a http://commons.wikimedia.org/wiki/File:Big_Burrs.jpg (6)
Figure 1.1b http://en.wikipedia.org/wiki/File:Velcro.jpg (1)
Figure 1.3a http://commons.wikimedia.org/wiki/File:Dead_ivy_vines_cling_to_tree.jpg (6)
Figure 1.3b http://en.wikipedia.org/wiki/File:Ancient_Egypt_rope_manufacture.jpg (7)
Figure 1.4a http://commons.wikimedia.org/wiki/File:Wasp_March_2008-9.jpg (1)
Figure 1.4b http://commons.wikimedia.org/wiki/File:Case_%C3%A0_la_chefferie_de_Bana.jpg (2)
Figure 1.5a http://en.wikipedia.org/wiki/File:Portuguese_Man_O%27_War_Miami_March_2008.jpg (Scott Sonnenberg) (8)
Figure 1.5b http://en.wikipedia.org/wiki/File:Kon-Tiki.jpg (3)
Figure 1.6a http://commons.wikimedia.org/wiki/File:Leonardo_Design_for_a_Flying_Machine,_c._1488.jpg (7)
Figure 1.6b http://en.wikipedia.org/wiki/File:Design_for_a_Flying_Machine.jpg (9)

Chapter 2
Figure 2.3 http://commons.wikimedia.org/wiki/File:PvT_3D_plot_-_water.png (1)
Figure 2.8 http://commons.wikimedia.org/wiki/File:Dewy_spider_web.jpg (1)
Figure 2.9 http://commons.wikimedia.org/wiki/File:Mangrove.jpg (2)
Figure 2.10a http://en.wikipedia.org/wiki/File:Big-eared-townsend-fledermaus.jpg (4)
Figure 2.10b http://commons.wikimedia.org/wiki/File:Green_Woodpecker_wing.jpg (Andreas Trepte) (5)
Figure 2.11 http://commons.wikimedia.org/wiki/File:US_Navy_050709-N-0000B-004_Hurricane_Dennis_batters_palm_trees_and_floods_parts_of_Naval_Air_Station_%28NAS%29_Key_West%5Ersquo,s_Truman_Annex.jpg (4)

Figure Attributions

Chapter 3
Figure 3.4 http://commons.wikimedia.org/wiki/File:Monkey_Wrench_%28PSF%29.png (2)
Figure 3.14 http://commons.wikimedia.org/wiki/File:3StageCreep.svg (12)
Figure 3.15 http://commons.wikimedia.org/wiki/File:Creep.svg (12)
Figure 3.16 http://commons.wikimedia.org/wiki/File:StressRelaxation.svg (12)
Figure 3.18 http://commons.wikimedia.org/wiki/File:Kelvin_Voigt_diagram.svg (2)
Figure 3.20 http://commons.wikimedia.org/wiki/File:Maxwell_diagram.svg (2)
Figure 3.22 http://commons.wikimedia.org/wiki/File:Kelvin_Voigt_diagram.svg (2)

Chapter 4
Figure 4.4 http://en.wikipedia.org/wiki/File:Von_Koch_curve.gif (2)
Figure 4.5 http://commons.wikimedia.org/wiki/File:Maple_Tree_Fractal_Branch_1.jpg (8)
Figure 4.11 http://commons.wikimedia.org/wiki/File:DOE_org_crart.PNG (4)
Figure 4.13 http://commons.wikimedia.org/wiki/File:TritiumRootCross.png (2)

Chapter 5
Figure 5.1 http://www.openclipart.org/detail/105775 (11)
Figure 5.3 http://commons.wikimedia.org/wiki/File:SEM_blood_cells.jpg (4)
Figure 5.4 http://commons.wikimedia.org/wiki/File:Spawn_hatched_fish_2.JPG (2)
Figure 5.5 http://commons.wikimedia.org/wiki/File:Blade_grass.jpg (Sebastian Mackiewicz) (6)
Figure 5.6 http://en.wikipedia.org/wiki/File:Bullfrog_-_natures_pics.jpg (5)
Figure 5.7 http://commons.wikimedia.org/wiki/File:Sir_Grantley_Adams_Int_Airport,_Barbados-03.jpg (2)
Figure 5.8 http://commons.wikimedia.org/wiki/File:Ram_air_square.jpg (4)
Figure 5.9 http://commons.wikimedia.org/wiki/File:Wallops_Balloon_With_BESS_Payload_DSC00088.JPG (3)
Figure 5.10 http://commons.wikimedia.org/wiki/File:Solar_sail_tests.jpg (3)

Chapter 6
Figure 6.1a http://en.wikipedia.org/wiki/File:WPThermostat.jpg (1)
Figure 6.1b http://en.wikipedia.org/wiki/File:Unitload.jpg (2)
Figure 6.1c http://en.wikipedia.org/wiki/File:Bug_Sized_Spies,_US_air_force.jpg (4)
Figure 6.1d http://en.wikipedia.org/wiki/File:Polymerization_in_situ.jpg (2)
Figure 6.2 http://en.wikipedia.org/wiki/File:WPThermostat.jpg (1)
Figure 6.3 http://commons.wikimedia.org/wiki/File:Neuron.jpg (4)
Figure 6.4 http://commons.wikimedia.org/wiki/File:Tree_sap.jpg (Justin Smith) (8)
Figure 6.5 http://en.wikipedia.org/wiki/File:Wound_healing_phases.png (2)
Figure 6.6 http://en.wikipedia.org/wiki/File:Polymerization_in_situ.jpg (2)

Figure Attributions

Chapter 7
Figure 7.1 (11)
Figure E.7.1 http://commons.wikimedia.org/wiki/File:Rhinoceros_beetle.jpg (10)

Appendix I
Figure A.1 http://commons.wikimedia.org/wiki/File:Scientific_classification.png (2)

Notes

(1) Permission is granted to copy, distribute, and/or modify this document under the terms of the GNU Free Documentation License, Version 1.2 or any later version published by the Free Software Foundation; with no Invariant Sections, no Front-Cover Texts, and no Back-Cover Texts. A copy of the license is included in the section entitled GNU Free Documentation License.

(2) I, the copyright holder of this work, release this work into the **public domain**. This applies worldwide.

(3) *This file is in the **public domain** because it was created by NASA. NASA copyright policy states that "NASA material is not protected by copyright **unless noted**."*

(4) *This work is in the **public domain** in the United States because it is a work of the United States Federal Government under the terms of* Title 17, Chapter 1, Section 105 of the US Code.

(5) This file is licensed under the Creative Commons Attribution-Share Alike 2.5 Generic license.

You are free:

to share—to copy, distribute, and transmit the work

to remix—to adapt the work

Under the following conditions:

- **attribution**—You must attribute the work in the manner specified by the author or licensor (but not in any way that suggests that they endorse you or your use of the work).
- **share alike**—If you alter, transform, or build upon this work, you may distribute the resulting work only under the same or similar license to this one.

(6) This file is licensed under the Creative Commons Attribution 2.0 Generic license.

You are free:

to share—to copy, distribute, and transmit the work

to remix—to adapt the work

Figure Attributions

Under the following conditions:

attribution—You must attribute the work in the manner specified by the author or licensor (but not in any way that suggests that they endorse you or your use of the work).

(7) *This image (or other media file) is in the **public domain** because its copyright has expired*.

(8) *This work is licensed under the Creative Commons Attribution-ShareAlike 3.0 License*.

(9) *This work is in the **public domain** in the United States, and those countries with a copyright term of life of the author plus **100** years or fewer.*

(10) This file is licensed under the Creative Commons Attribution 1.0 Generic license.

(11) Unless noted, content on this site is waived of all copyright and related or neighboring rights under the CC0 PD Dedication.

(12) This file is made available under the Creative Commons CC0 1.0 Universal Public Domain Dedication.

The person who associated a work with this deed has dedicated the work to the public domain by waiving all of his or her rights to the work worldwide under copyright law, including all related and neighboring rights, to the extent allowed by law. You can copy, modify, distribute and perform the work, even for commercial purposes, all without asking permission.

Index

Note: *f* indicates a figure; *t* indicates a table

A

Abstract design space, 12, 12*f*
Actin, 81
Active resin infusion systems, 113
Adaptationism, 22
Allometry/allometric scaling, 58–61
    log–log plot, 59*f*
Altshuller, Genrich, 122
Area-forming elements, 29, 29*t*
Artifacts, biological, 123–24
AskNature (www.asknature.org), 120
Autonomic reflexes, 110
Axial structures, 29, 32–33, 32*f*
    boundary conditions, 33
    free-body diagram of, 34, 34*f*
    loads on, 33

B

Beams
    bending of, 96–99, 96*f*
    LFE and, 30
Bending shape factors, 75–77
Beneficial traits, design in nature and, 19–20
Bimetallic thermostat, 109, 109*f*
Bioinspiration & Biomimetics, 121
Bio-inspired engineering (BiE)
    defined, 2–3
    forward design process in, 117–19, 118*f*
    inverse design process in, 118, 119, 122–26
    Mestral, George De, 1–2, 3
    resources for, 119–22
        scholarly journals, 120–22
        web databases, 120
    Reverse Bio-Inspired Engineering Worksheet, 123–25
    rope, 4, 4*f*
    shelter, 5, 5*f*
    transportation, 5, 5*f*
    See also Engineering

Biological classification, 131–32, 131*f*
Biological materials, 77–81
    building blocks, 80–81
    comparative properties of
        modulus *vs.* density, 143*f*
        modulus *vs.* strength, 145*f*
        strength *vs.* density, 144*f*
    features of, 77
    mechanical properties of, 81*t*
    relative distribution in humans, 81*t*
    structural hierarchy, 78–79
        organizational, 78*f*
        in plant root, 80*f*
        in systems biology, 79*f*
Biological smartness, 109–10
Biology
    self-healing in, 111–13
    vs. engineering, 6–7
Biomimetics, 3
Biomimicry, 3
Biomimicry Institute, 3
"Bionic," 2–3
Body loads, 27
Boundary conditions, 30–31, 31*t*
    axial structures, 33
Brainstorming, 118–19
Branch, 63*f*
    cross-sectional areas of, 64
Building blocks, 80–81
Burdock seeds, 1, 2*f*

C

Cables
    hanging, 99–104, 100*f*
        compliant (nonlinear) model, 102
        stiff (linear) model, 102
    LFE and, 30
"Catalog of biological solutions," 119
Catenary, 102

Cauchy's stress principle, 35
CDOF. *See* Coordinate degrees of freedom (CDOF)
Ceramics, 80
Chitin, 81
Closed cross section, thin-walled shafts of, 70–71, 70f
Coaxial loads, 33
Coefficient of viscosity, 48
Collagen, 80
Column buckling, 91–96
    compliant, 93–96, 93f
    stiff (linear) model of, 91–93, 91f
Combined shape factor, 77
Compliant (nonlinear) model
    of column buckling, 93–96, 93f
    of hanging cable, 102
Compliant structures
    membrane, 85–86
    in nature and engineering, 87f–90f
    paradigm in nature, 85–87, 86f
    response analysis
        beam bending, 96–99
        column buckling, 91–96
        hanging cable, 99–104
Composites, 80
Compressive force, 33
Concentrated force, 33
Configuration
    application, 32–33, 32f–34f
    structural analysis and, 32
Conjugate stress–strain pairs, 44
Constitution, structural analysis and, 32
Constitutive equation, 41
Constitutive models, 41–52
    linear elasticity, 42–43, 42f
    linear viscoelastic constitutive models, 47–50, 47f–50f
    linear viscoelasticity, 45–47, 45f–47f
    nonlinear elasticity, 43–44, 43f–44f
    practical considerations, 41
    Standard Linear Solid, 50–52, 50f–51f
Constrained optimization problem, 14, 14f
Constructal theory, 122
Continuum hypothesis, 32
Coordinate axes, 11, 11f
Coordinate degrees of freedom (CDOF), 11

Creep, 45–46, 46f
    mechanisms of, 47
    rupture, 47
Creep compliance, 48
Creep rupture, 47
Cylinder, surface area of, 65–66

D

Da Vinci, Leonardo, 6
DDOF. *See* Design degrees of freedom (DDOF)
Deciding, action of, 107
Deformation
    application, 38–41, 40f
    structural analysis and, 32
De Mestral, George, 119
    historical perspectives, 1–2, 3
    "Velcro"®, 1, 2f, 4f
Dendritic tree, of neuron, 110, 111f
Density, of natural materials
    vs. modulus, 143f
    vs. strength, 144f
Design, engineering, 9–23
    optimal design, 13–18, 14f–18f
    process, 9–13, 10f–13f
        forward *vs.* inverse design, 10–11, 10f
        total design, 10
    rules, 13
    See also Design process, in bio-inspired engineering
Design degrees of freedom (DDOF), 12, 13, 42
Design in nature, 18–23
    beneficial traits and, 19–20
    evolution and optimality, 22–23
    process, 18–22, 20f–22f
    rules, 19–20, 20f–22f
Design process, in bio-inspired engineering
    forward, 117–19, 118f
    inverse, 118, 119, 122–26
Design solution space, 11
Design space, 12, 12f
    abstract, 12, 12f
    constrained by limits on one of DDOF, 12–13, 13f
Deviatoric effects, 42
Dilatational effects, 42
Dynamic analysis, 32
Dynamic loads
    steady-state loads, 28
    transient loads, 28

Index

E

EAPs. *See* Electroactive polymers (EAPs)
Edwards, James, Dr., 120
Effector, 110
Elasticity, atomistic basis for, 42–43
Elastic modulus, 32
Elastomers, 80
Electroactive polymers (EAPs), 108
Electrorheological fluid (ERF), 108
Encyclopedia of Life (www.eol.org), 120
Engineering
    bio-inspired engineering (BiE)
        (*See* Bio-inspired engineering (BiE))
    design, 9–23 (*See also* Design, engineering)
    self-healing in, 111–13
    structures and materials in, 27–52 (*See also* Structures and materials)
    vs. biology, 6–7
Environmental pressures, 58
Equations of constraint, boundary conditions and, 30
Equations of motion, 32, 41
Equilibrium
    application, 34–38, 34f–38f
    equations, 41
    structural analysis and, 32
ERF. *See* Electrorheological fluid (ERF)
Euler, Leonard, 91
Euler column buckling, 91, 91f
Evolution, 19
    optimality and, 22–23

F

FAPs. *See* Fixed action patterns (FAPs)
FBD. *See* Free-body diagram (FBD)
Fick's first law, 65
Fish, Frank, Dr., 119
Fixed action patterns (FAPs), 110
Forward design process, 10
    in bio-inspired engineering, 117–19, 118f
    vs. inverse design, 10–11, 10f
Fractal scaling, 62, 62f, 63f
Free-body diagram (FBD)
    of axial structure, 34, 34f
Functional analysis, 118

G

Geometric scaling, 58, 58f
Gould, Stephen, 22

Gravity, 58
    effects on scaling, 62

H

Hanging cable, 99–104, 100f
    compliant (nonlinear) model, 102
    stiff (linear) model, 102
HD. *See* Hierarchical design (HD)
Heuristic, 13
Hierarchical design (HD)
    design in nature and, 19
Hierarchical organization., 78f
Hooke, Robert, 42
"Hookean" material, 42
Human-designed sailcraft, 5, 5f
Hyperelastic material, 43

I

Incompressible materials, 44
Integration center, 110
Intelligence *vs.* smartness, 109
International Journal of Bio-Inspired Computation, 121–22
International Journal of Design & Nature and Ecodynamics, 120–21
Inter-neurons reflexes, 110
Inverse design process, 10
    in bio-inspired engineering, 118, 119, 122–26
    vs. forward design, 10–11, 10f
Inverse mechanics problem, 133–34
Isometry/isometric scaling, 58–59
    log–log plot, 59f
    See also Allometry/allometric scaling

J

Journal of Biomimetics, Biomaterials, and Tissue Engineering, 121
Just good enough design, 20, 20f, 21f

K

Kelvin–Voigt Solid, 47–49, 48f–49f
Keratin, 81
Kinematic relations, 39, 41
Koch "snowflake," geometry of, 62f

L

Lewontin, Richard, 22
LFEs. *See* Line-forming elements (LFEs)

153

Index

Life cycle design, 10
Linear elasticity, 42–43, 42f
Linear elastic material, loading–unloading curve for, 42, 42f
Linear viscoelastic constitutive models, 47–50, 47f–50f
   Kelvin–Voigt Solid, 47–49, 48f–49f
   Maxwell fluid, 49–50, 49f–50f
Linear viscoelasticity, 45–47, 45f–47f
Line-forming elements (LFEs), 29–30, 29t
   beams, 30
   cables, 30
   rods, 29–30
Live loads, 27–28
LM. *See* Low-modulus materials (LM)
Loads, structural, 27–28, 28f
   on axial structure, 33
   body loads, 27
   dynamic loads, 28
   orientation of, 28, 28f
   static, 27–28
   surface loads, 27
Long-term modulus, 51
Low-modulus materials (LM), design in nature and, 19–20, 20f

M

Magnetorheological fluid (MRF), 108
Mass density, 32
Maxwell fluid, 49–50, 49f–50f
Membrane, 85–86, 86f
Modulus, of natural materials
   vs. density, 143f
   vs. strength, 145f
Moment–twist (gradient) relation, 66–67
Monosynaptic reflexes, 110
Mooney–Rivlin model, 45
Motor neuron, 110
MRF. *See* Magnetorheological fluid (MRF)

N

Natural materials. *See* Biological materials
Natural selection, 19
Nature
   compliant paradigm in, 85–87
   configuration, 57–62
      allometry, 58–61
      fractal scaling, 62, 62f, 63f
      scaling, gravity effects on, 62
      size (background), 57–58, 58f
   environmental pressures, factors of, 58
   self-healing, 111–13
   shape and structure, 63–77
      bending shape factors, 75–77
      bifurcation branching, 63f
      combined shape factor, 77
      thin-walled shafts of closed cross section, 70–71, 70f
      thin-walled shafts of open cross section, 72–73, 72f
      thin-walled torsion structures, 69–70
      torsion revisited, 66–69
      torsion shape factor, 73–75
Nature Materials, 121
Nerve, 110
Neurons, 110–11, 111f
Nickel–titanium alloys (nitinol), 108
Nonlinear elasticity, 43–44, 43f–44f
Nonlinear elastic material, loading–unloading curve for, 43, 43f
Nylon, 3

O

Open cross section, thin-walled shafts of, 72–73, 72f
Optimal design, engineering, 13–18, 14f–18f
Optimality, evolution and, 22–23
Optimal solution, 14

P

Partition of structural function (PSF), design in nature and, 19, 21f
Physical coordinate space, 11, 11f
Piezoelectrics, 108
Plates, SFE and, 30
Polar moment of inertia, 33, 66
Polymers, 80
Polysynaptic reflexes, 110
Prismatic structures, 33
Proteins, 80
PSF. *See* partition of structural function (PSF)

Q

Quasi-static loads, 28

R

Random genetic drift, 19
Receptor, 110
Recovery, 46, 46f

Reflex arcs, 110
  components of, 110
Reflexes, 110
  autonomic, 110
  somatic, 110
Relaxation
  modulus, 49
  stress, 47, 47f
"Resin infusion," 112, 112f, 113, 113f
Responding, action of, 107
Response analysis, compliant structures
  beam bending, 96–99
  column buckling, 91–96
  hanging cable, 99–104
Reverse Bio-Inspired Engineering Worksheet, 123–25, 135–39
Reverse design process. *See* Inverse design process
Reverse engineering, 17
  worksheet, 123–25, 135–39
Rheological fluids, 108
Rods, LFE and, 29–30
Rope in biology, 4, 4f
Rotation, boundary conditions and, 30

S

Scaling
  concept of, 58
  fractal, 62, 62f, 63f
  geometric, 58f
  gravity effects on, 62
Schmitt, Otto, 3
SDOF. *See* Single degree of freedom (SDOF)
Self-healing, 111–13. *See also* Wound healing
Self-similar geometry, 62
Sensory neuron, 110
SFEs. *See* Surface-forming elements (SFEs)
Shape factors
  bending, 75–77
  combined, 77
  torsion, 73–75
Shape memory alloys/polymers (SMA/SMP), 108
Shear flow, 69–70
Shearing strain, 41
Shear modulus, 66
Shear strain–displacement relation, 66
Shells, SFE and, 30

Shelter, 5, 5f
Single degree of freedom (SDOF), 133, 133f
Smart materials, 108
  classification, 108
Smartness
  biological, 109–10
  defined, 109
  vs. intelligence, 109
Smart structural system, 107–13
  bimetallic thermostat, 109, 109f
  defined, 107
  overview, 107
  range of complexity, 108f
  three actions of, 107
SMA/SMP. *See* Shape memory alloys/polymers (SMA/SMP)
Somatic reflexes, 110
Space, 12
St. Venant's principle, 35
Standard Linear Solid, 50–52, 50f–51f
Static analysis, 32
Steady-state loads, 28
Stiff (linear) model
  of column buckling, 91–93, 91f
  of hanging cable, 102
Stiffness, 31
Strain, defined, 38–39
Strain–displacement relation, 39
Strain energy density function, 43
Strength, 31
  of natural materials
    vs. density, 144f
    vs. modulus, 145f
Strength failures, design in nature and, 19–20, 20f
Stress
  distribution, 35
  relaxation, 47, 47f
  state, 35
Stress–strain relations, 32
Stretch, concept of, 44, 44f
Structural analysis, 31–32
  configuration, 32
  constitution, 32
  deformation, 32
  equilibrium, 32
Structural compliance, 85. *See also* Compliant structures

Index

Structural form
    line-forming elements, 29–30, 29t
    surface-forming elements, 29, 29t, 30
    taxonomy of, 29–30, 29t
Structural hierarchy, design in nature and, 19
Structural loads. *See* Loads, structural
Structures and materials
    boundary conditions, 30–31, 31t
    constitutive models, 41–52 (*See also* Constitutive models)
    structural analysis, methods of, 31–32
    structural form, 29–30, 29t (*See also* Structural form)
    structural loads, 27–28, 28f
Surface area, of cylinder, 65–66
Surface-forming elements (SFEs), 29, 29t, 30
    area-forming elements, 29, 29t
    volume-forming elements, 29, 29t
Surface loads, 27
Systems biology, hierarchy in, 79f

T

Taxonomy, 131, 131f
Tensile force, 33
Tensile strength, 32
Theory of Inventive Problem Solving (TIPS), 122
Thin solid rectangular shaft, cross section of, 72–73, 72f
Thin-walled shafts
    of closed cross section, 70–71, 70f
    of open cross section, 72–73, 72f
Thin-walled torsion structures, 69–70
Three-horned rhinoceros beetle, 123f
Time–temperature superposition, 47
TIPS. *See* Theory of Inventive Problem Solving (TIPS)
Torque, 33, 34f

Torsion
    stiffness, 67
Torsion, revisited, 66–69
Torsion constant, 71
Torsion shape factor, 73–75
Torsion structure, 33, 33f
Total design, considerations for, 10
Transient loads, 28
Translation, boundary conditions and, 30
Transportation, 5, 5f
Tree of Life (www.tolweb.org/tree), 120
TRIZ, 118–19, 122. *See also* Theory of Inventive Problem Solving (TIPS)
"Tubercle Technology," 119

U

Uniform stress (US), design in nature and, 20
US. *See* Uniform stress (US)

V

"Velcro"®, 1, 2f, 4f, 119
    De Mestral, George, 1–2, 3
Vestibulo-ocular reflex (VOR), 111
Vines, 4, 4f
Viscoelastic materials, characteristics, 45–47
Volume-forming elements, 29, 29t
von Karman, Theodore, 6
VOR. *See* Vestibulo-ocular reflex (VOR)

W

Web databases, for bio-inspired engineering, 120
Whale Power, 119
Wound healing
    stages of, 112–13, 113f
    in a tree, 112, 112f
    See also Self-healing